DRONES OVER THE BACK FORTY

A Midwest Farmer's Guide to Seeing What Your Eyes Simply Can't

Adam Miller

Black Swamp Press
Beavercreek, Ohio

Published by Black Swamp Press, Beavercreek, Ohio

ISBN: 979-8-9961617-0-6

First Edition: May 2026

Printed in the United States of America

Disclaimer: This book is for informational and educational purposes only. Drone operations are governed by FAA regulations subject to change. Agricultural decisions should be made in consultation with qualified agronomists and local Extension professionals. Mention of specific products does not constitute endorsement.

For the farmers.

The ones who wake before sunrise and work past dark. Who carry the weight of weather, markets, and debt on shoulders that rarely complain. Who feed a nation that rarely thinks about where its food comes from. Thank you, sincerely, for what you do. I hope this book adds to the immense knowledge you already have.

Table of Contents

"The study of tools as well as of books should have a place."

— Rutherford B. Hayes

"For the farmer is the only man in our economy who buys everything he buys at retail, sells everything he sells at wholesale, and pays the freight both ways."

— John F. Kennedy

"To sit back and let fate play its hand out and never influence it is not the way man was meant to operate."

— John Glenn

Introduction

Let me tell you what this book is not.

It's not a tech manual written by someone who's never had dirt under their fingernails. It's not a sales pitch from a drone company hoping you'll buy the most expensive sensor on the shelf. And it's not an academic textbook that explains the physics of light in language that makes your eyes glaze over.

This book is a straight-talking, practical guide for working farmers who want to know one thing: can a drone actually save me money and help me make better decisions on my operation?

The honest answer is: for most Midwest row crop operations above a couple hundred acres, yes. But not in the way the marketing brochures suggest, and not without understanding what you're looking at.

Why I Wrote This

I grew up in Napoleon, Ohio—a small town in the heart of northwest Ohio farm country where the flat, dark soil stretches to the horizon and the rhythm of planting and harvest shapes everything. I didn't grow up on a farm, but I grew up in farming. Friends, neighbors, and extended family worked the land. In high school, I spent my summers baling hay for a local farm, fitting the work in between baseball games and football practice. It was honest, hot, physical labor—the kind that gives you a permanent respect for the people who do it year-round.

In 2006, I enlisted in the United States Air Force and spent the next twenty years as a geospatial intelligence analyst—one of the people who stares at imagery from satellites, high-altitude aircraft, and unmanned platforms and turns what they see into actionable intelligence. My specialty is exactly the technology that agricultural drones are built on: multispectral sensors, thermal infrared imaging, electro-optical cameras, and the analytical methods that make raw sensor data meaningful. The platforms were different—U-2s, Global Hawks, Predators, Reapers—but the fundamental science is identical to what a DJI Mavic does over a cornfield.

When I came home to Ohio, I noticed something: the same remote sensing technology I'd spent my career mastering was beginning to appear over those same farm fields. Farmers were being sold drone technology by people who understood the gadgets but didn't understand farming, or by agronomists who understood the crops but couldn't explain the sensors. Nobody was bridging that gap in plain language. So I wrote it.

Who This Book Is For

This book is written for multigenerational family farmers in the Midwest—primarily corn, soybean, and wheat operations ranging from 200 to 2,000 acres. If you're operating in Ohio, Indiana, Illinois, Iowa, Michigan, or anywhere in the heart of the Corn Belt, the examples, economics, and agronomic references in these pages were written with your fields in mind.

You're probably somewhere between 35 and 70 years old. You've got strong agronomic intuition built over decades of walking your fields. You're skeptical of anything that doesn't prove its value in dollars saved or bushels gained. And you've been hearing about drones for a few years now, wondering whether it's time to pay attention.

It is. And this book will show you exactly why—and exactly how.

A Note on Honesty

This book mentions specific products by name—DJI drones, Pix4D software, MicaSense sensors—because farmers make purchasing decisions and they need real information to do it. I have no financial relationship with any manufacturer, retailer, or service provider mentioned in these pages. When I recommend a product, it's because I've used it, tested it, or seen it perform in real field conditions. You deserve straight answers. That's what this book delivers.

How to Use This Book

This book is designed to work two ways: straight through, or as a reference you pick up when you need it.

If You're Brand New to Drones

Start at Chapter 1 and read through Chapter 11. By the end, you'll have your Part 107 license, a working knowledge of what drone maps can show you, and a 12-week action plan for your first season. Then jump to Chapter 19 (Common Mistakes) before you fly your first real mission.

If You Already Own a Drone

Jump directly to the chapter that matches your need: Chapter 12 for nitrogen management, Chapter 13 for stand counts and scouting, Chapter 14 for in-season monitoring, Chapter 15 for disease detection, or Chapter 16 for variable-rate prescriptions. Each chapter stands on its own.

If You're Evaluating Whether to Invest

Read Chapters 1 and 2 for the honest cost-benefit case. Then skip to Chapter 7 for the RGB-vs.-multispectral investment decision—it includes break-even calculations for different acreage levels.

A Note on Callout Boxes

Throughout the book you'll find five types of highlighted boxes: KEY INSIGHT, PRACTICAL EXAMPLE, COST ANALYSIS, CAUTION, and PRO TIP. Each signals the kind of information inside so you can read straight through or scan for what you need.

PART 1 — Why This Matters (And When It Doesn't)

Chapters 1–3

Before you spend a dollar on equipment, you need to know whether drone technology makes financial sense for your operation. Part I makes the honest case—what a $1,200–$1,500 investment can realistically deliver, what it can't, and the legal requirements you need to satisfy before your first flight. Chapter 1 establishes what decades of field experience have already taught you and where the gaps are. Chapter 2 cuts through the marketing noise to show exactly what equipment you need (and what you don't). Chapter 3 walks you through the FAA Part 107 certification and the regulations that govern every flight. By the end of Part I, you'll know whether you're in or out—and you'll have a clear picture of the entry cost.

Chapter 1: What 30 Years of Walking Your Fields Taught You—And What It Couldn't

You know your fields. You know where it stays wet in May, where the tile lines run, which spots always seem to yellow up first. That knowledge took decades to build. This book isn't here to replace it.

What You Already Know

If you've been farming for twenty or thirty years in the Midwest, you carry around a mental map of your operation that no satellite image can match. You remember the summer of '12 when the drought hit and that back forty never fully recovered. You know which field drains fastest after a heavy rain and which one will still have standing water three days later. You've watched your father or grandfather work this same ground, and some of what you know came from them—passed down through observation, conversation, and the hard lessons of seasons that didn't go as planned.

This accumulated knowledge is irreplaceable. No drone, no sensor, no algorithm developed in Silicon Valley can tell you that the northwest corner of Field 7 always seems to have compaction issues because that's where your neighbor used to turn his equipment around before you bought the property in 1998. Technology doesn't know that the creek flooding in 2008 deposited a layer of sand along the east edge of your best bean field, changing the soil characteristics in ways that still show up in your yields today.

The average age of a U.S. farm producer in 2022 was 58.1 years, according to the USDA Census of Agriculture—up from 57.5 in 2017, continuing a long-term trend.[1] These producers have been farming an average of 23.4 years.[2] That's over two decades of accumulated knowledge about soil, weather patterns, crop behavior, and the thousand small variables that determine whether a season ends in profit or loss.

SNAPSHOT: U.S. FARM PRODUCERS (2022 CENSUS OF AGRICULTURE)

Average age of all producers: 58.1 years

Average years of farming experience: 23.4 years

Producers age 65 and older: 38% of all producers

Family-owned and operated farms: 95% of all U.S. farms

Source: USDA National Agricultural Statistics Service, 2022 Census of Agriculture

Family-owned and operated farms account for 95% of all U.S. farms, operating 84% of land in farms.[3] These aren't corporate operations running on spreadsheets and shareholder reports. These are operations where knowledge passes from one generation to the next, where the person making decisions in the cab of the combine is the same person who planted that seed, walked those rows during the growing season, and will be there when the banker wants to discuss operating loans in February.

What You Can't See

Here's the thing: you can't be everywhere at once.

During planting, you're focused on getting seed in the ground while weather windows allow. During the growing season, you're managing a dozen tasks simultaneously – checking fields, maintaining equipment, dealing with suppliers, keeping up with paperwork. You can't walk every acre every week, especially not during the critical growth stages when problems are most likely to develop and most important to catch early.

You can't see what's happening in the back forty while you're planting the front eighty. You can't spot the first signs of nitrogen stress before it shows up to the naked eye—by the time corn starts "firing" with yellow lower leaves, you've already lost yield potential. You can't detect the beginnings of a disease outbreak in the middle of a 160-acre field from the end rows.

And you can't easily see patterns that only become apparent from above. That streaking pattern that follows your planter passes? It's obvious from 250 feet up, but nearly invisible when you're standing in the field. The circular stress pattern that indicates a compaction problem from where the grain cart turns? You might drive past it for years without realizing it's there.

KEY INSIGHT

Drone technology doesn't replace what you know – it shows you what you can't see. Your decades of experience tell you WHY something is happening. The aerial view shows you WHERE it's happening, often weeks before it becomes visible from the ground.

The Gap Between Knowing and Seeing

Consider a common scenario: you know from experience that a certain area of your field tends to be less productive. Maybe it's a low spot that stays wet, or a knob where the topsoil is thin. You've adjusted your expectations for that area over the years, perhaps even thought about tiling or soil amendments.

But do you know exactly where that problem zone starts and stops? Do you know if it's 5 acres or 15 acres? Do you know if it's getting better or worse over time? Do you know if your variable-rate fertilizer prescription is actually matching the boundaries of the problem, or if you're wasting inputs on the margins?

This is where the combination of farmer knowledge and aerial technology becomes powerful. You provide the "why"—decades of understanding about your soil, your management practices, your local conditions. The technology provides the "where" and "how much"—precise boundaries, quantified differences, documented changes over time.

The Technology Gap in Adoption

Despite the potential benefits, precision agriculture adoption remains limited on many farms. According to the U.S. Government Accountability Office, only 27% of U.S. farms or ranches used precision agriculture practices to manage crops or livestock as of 2023.[4] This is only a modest increase from 25% reported in the 2021 USDA Farm Computer Usage and Ownership Survey.

The adoption picture varies dramatically by farm size and location. In the leading row-crop states – North Dakota, Nebraska, Iowa, South Dakota, and Illinois – more than half of producers report using some form of precision agriculture.[5] These five states account for nearly half of all U.S. corn and soybean cash receipts. But nationwide, most farms haven't yet adopted these technologies.

Table 1.1: Precision Agriculture Adoption by State (2023)

State	% of Farms Using Precision Ag
North Dakota	57%
Nebraska	55%
Iowa	54%
South Dakota	53%
Illinois	51%
Ohio	40% (estimated)
Indiana	40% (estimated)
U.S. Average	27%
Source: USDA NASS, Farm Computer Usage and Ownership Report (2023); U.S. GAO Report GAO-24-105962	

When you break down adoption by specific technologies, the picture becomes more nuanced. Guidance and autosteering systems – which help reduce overlap and improve planting accuracy – have seen the widest adoption, used by 52% of midsize farms and 70% of large-scale crop-producing farms in 2023.[6] These technologies offer immediate, tangible benefits that are easy to see and measure.

But aerial imagery – from drones, aircraft, or satellites – remains less common. According to the USDA Economic Research Service, aerial imagery was used on only about 7% of corn acres in 2016 and 10% of soybean acres in 2018.[7] While these numbers have likely increased since then, aerial monitoring remains an underutilized tool on most farms.

Why the Hesitation?

If you're reading this book and haven't yet invested in drone technology, you're in good company. There are legitimate reasons why many experienced farmers have held back:

- Cost concerns. Precision agriculture technologies can require significant upfront investment. The GAO found that high costs remain a primary barrier to adoption, particularly for smaller operations.[8]
- Complexity. Learning new technology takes time—time that's in short supply during the growing season. And if you're not confident you can interpret the data, what's the point of collecting it?
- Uncertain ROI. Unlike a new planter or a tile drainage project, the return on precision agriculture investment can be hard to quantify. How do you measure the value of information?
- Skepticism. You've seen technology fads come and go. Not every "revolutionary" tool turns out to be worth the hype. A healthy skepticism has probably saved you money over the years.
- Connectivity issues. Many precision agriculture tools require internet connectivity, and rural broadband remains limited. Farms with internet access rose from 75% in 2017 to 79% in 2022, but that still leaves more than one in five farms without reliable access.[9]

These are all valid concerns. This book won't try to convince you that technology is always the answer or that you need the most expensive equipment on the market. Instead, the goal is to give you enough practical information to make an informed decision about whether drone technology makes sense for your specific operation.

The Real Question

This isn't about whether drones are "cool" or whether you need the latest gadget. The real question is simple: Can a $1,500 piece of equipment and a few hours of flight time save you money and help you make better decisions?

For some operations, the answer is a clear yes. If you're managing 500+ acres of row crops, dealing with variable soil types, and already making decisions about variable-rate applications, the information from even basic drone imagery can pay for itself quickly. A single avoided misapplication of nitrogen – catching a problem early instead of blanket-treating an entire field – can easily exceed the cost of entry-level equipment.

For other operations, the answer is "not yet." If you're farming under 200 acres, have relatively uniform fields, and aren't in a position to act on variable-rate information anyway, the investment might not make sense right now. There's no shame in waiting until the technology proves itself further or until your operation grows to a scale where the benefits become clearer.

This book will help you figure out which category you're in – and if you decide to move forward, how to do it in a way that maximizes value while minimizing risk and frustration.

What the Research Shows

Before dismissing precision agriculture as expensive gadgetry, consider what university research and real-world applications have demonstrated:

Nitrogen management offers the clearest financial benefit. Studies from the University of Nebraska and other land-grant institutions have shown that variable-rate nitrogen application—informed by remote sensing data—can reduce nitrogen use by 15–20% while maintaining or improving yields.[10] At current fertilizer prices, that translates to meaningful savings on every acre.

Remote sensing-based nitrogen management in one Colorado study reduced total nitrogen leaching losses by 43–51% compared to standard farmer practices.[11] That's not just an economic benefit—it's an environmental benefit that increasingly matters for market access and regulatory compliance.

According to research from the USDA Economic Research Service, farmers who effectively adopt precision agriculture technologies see average corn yield improvements of 5–10 bushels per acre.[12] At $4.50 per bushel, that's an additional $22.50–$45 per acre in revenue – results that vary based on field variability and how actively data is used for decision-making.

DOING THE MATH: A SIMPLE ROI EXAMPLE

500-acre corn operation, typical Western Ohio conditions:

- *Entry-level drone equipment: $1,500*
- *Conservative N savings (10%): 15 lbs/acre x $0.50/lb = $7.50/acre*
- *Total potential N savings: $3,750/year*
- *Plus: Early problem detection, stand count verification, drainage assessment*

Payback period: Potentially less than six months of one growing season.

Note: Actual results vary based on field variability, management practices, and how the data is used for decision-making.

A Different Kind of Technology Book

You may have noticed that most books and articles about precision agriculture are written by engineers, data scientists, or academics. They're full of equations, technical specifications, and theoretical discussions that assume you have time to become an expert in remote sensing before you can use it.

This book takes a different approach. It's written for farmers who already know how to farm – who have the agronomic knowledge and practical experience but want to understand how aerial monitoring can enhance what they already do. The goal is not to turn you into a GIS specialist or a drone technician. The goal is to help you understand enough about the technology to use it effectively, interpret the results confidently, and make better decisions for your operation.

Throughout this book, we'll focus on practical applications specific to Midwest row crop production. The examples will use corn and soybeans in Ohio clay soil, not California almonds or Florida oranges. The equipment recommendations will be based on what's actually available and affordable, not the latest research prototype. And we'll be honest about limitations – what drones can and can't do, when the technology makes sense and when it doesn't.

How This Book Is Organized

This book is structured to take you from "curious but skeptical" to

"confident and capable" in a logical progression:

Part One (Chapters 1–3) helps you decide if drone technology makes sense for your operation. We'll cover equipment options at various price points, legal requirements, and how to realistically assess potential returns.

Part Two (Chapters 4–7) explains the science in practical terms. You'll learn how plants "talk" through the light they reflect, what vegetation indices actually measure, and when different types of cameras and sensors make sense.

Part Three (Chapters 8–11) covers flight operations—practical guidance on flight planning, software options, how to interpret the maps you create, and a 12-week first-season action plan.

Part Four (Chapters 12–16) connects data to decisions. This is where the rubber meets the road—using aerial information for nitrogen management, scouting, in-season monitoring, problem diagnosis, and variable-rate prescriptions.

Part Five (Chapters 17–19) covers the practical realities of running a drone program on a working farm—when to fly yourself versus hiring a service provider, integrating drone data with your existing precision-agriculture tools, and the most common mistakes new operators make.

Appendixes provide reference materials you can use in the field – quick reference cards, equipment recommendations by budget, software tutorials, regional resources, and worksheets for tracking ROI.

You don't have to read this book cover to cover. If you're primarily interested in nitrogen management, skip to Chapter 12. If you just want to know what equipment to buy, Chapter 2 has you covered. Use the index and appendices to find specific information when you need it.

Chapter Summary

- Your decades of farming experience provide irreplaceable knowledge about your operation – knowledge that no technology can duplicate.
- However, you can't be everywhere at once, and you can't see what's happening below the canopy or across hundreds of acres simultaneously.
- Only 27% of U.S. farms currently use precision agriculture practices, with adoption much higher in major row-crop states.
- Cost, complexity, uncertain ROI, and connectivity issues are legitimate barriers to adoption.
- Research demonstrates meaningful benefits from precision agriculture, including nitrogen savings of 15–20% and yield increases of 5–10 bushels per acre in corn, per USDA ERS research.
- The goal of this book is to help you decide if drone technology makes sense for your operation, and if so, how to implement it effectively.

Self-Assessment: Is Drone Technology Right for My Operation?

Use this checklist to help determine if drone-based crop monitoring might benefit your farm. The more statements you agree with, the more likely you are to see positive returns from the investment.

OPERATION CHARACTERISTICS

- [ ] I manage 200+ acres of row crops (corn, soybeans, wheat)
- [ ] My fields have significant variability in soil type, drainage, or productivity
- [ ] I already use or plan to use variable-rate application technology
- [ ] I have difficulty scouting all my acres as often as I'd like during the season
- [ ] I've experienced yield losses from problems I discovered too late
- [ ] I use yield monitors and have yield maps from previous seasons
- [ ] Nitrogen fertilizer represents a significant portion of my input costs
- [ ] I have time (or family members/employees who have time) to learn new technology
- [ ] I have reliable cellular or internet connectivity in most of my fields
- [ ] I'm interested in documenting field conditions for insurance or records

Scoring Guide

7–10 checks: Strong candidate for drone technology. The potential benefits likely outweigh the costs and learning curve.

4–6 checks: Moderate candidate. Consider starting with basic equipment or using a service provider to test the value before investing.

0–3 checks: May want to wait. The technology might not provide enough benefit yet for your current operation. Revisit as your operation changes or technology becomes more accessible.

Chapter 1 Notes and References

1. USDA National Agricultural Statistics Service. "Farm Producers." 2022 Census of Agriculture Highlights, February 2024. https://www.nass.usda.gov/Publications/Highlights/2024/Census22_HL_FarmProducers_FINAL.pdf

2. Ibid.

3. USDA National Agricultural Statistics Service. "USDA releases 2022 Census of Agriculture data." News Release, February 13, 2024. https://www.nass.usda.gov/Newsroom/2024/02-13-2024.php

4. U.S. Government Accountability Office. "Precision Agriculture: Benefits and Challenges for Technology Adoption and Use."

5. GAO-24-105962, January 2024. https://www.gao.gov/products/gao-24-105962

6. Ibid.

7. USDA Economic Research Service. "Precision agriculture use increases with farm size and varies widely by technology." Charts of Note, December 2023. https://www.ers.usda.gov/data-products/charts-of-note/chart-detail?chartId=110550

8. USDA Economic Research Service. "Aerial imagery remains mostly grounded on U.S. farms." Charts of Note, February 2023. https://www.ers.usda.gov/data-products/charts-of-note/chart-detail?chartId=107207

9. U.S. Government Accountability Office. GAO-24-105962, 2024.

10. USDA National Agricultural Statistics Service. 2022 Census of Agriculture.

11. Thompson, L. and Puntel, L. "Optimize your Nitrogen Investment with Precision Agriculture Technologies." University of Nebraska-Lincoln CropWatch, November 2021. https://cropwatch.unl.edu/2021/optimize-your-nitrogen-investment-precision-agriculture-technologies/

12. Delgado, J.A. and Bausch, W.C. "Potential use of precision conservation techniques to reduce nitrate leaching in irrigated crops." Journal of Soil and Water Conservation, 60(6), 379–387. Study conducted in northeastern Colorado.

13. McFadden, J., Njuki, E., and Griffin, T. "Precision Agriculture in the Digital Era: Recent Adoption on U.S. Farms." USDA Economic Research Service, Economic Information Bulletin No. 248. Washington, DC: USDA ERS, February 2023. https://www.ers.usda.gov/publications/pub-details/?pubid=105893

Additional Resources

- USDA Census of Agriculture: https://www.nass.usda.gov/AgCensus/

- USDA Economic Research Service - Precision Agriculture: https://www.ers.usda.gov/publications/pubdetails?pubid=105893
- Ohio State University Extension: https://www.extension.osu.edu/
- Purdue University Precision Agriculture: https://www.ag.purdue.edu/digital-ag-resources/

Chapter 2: The $1,500 Entry Point—What You Actually Need vs. What They're Selling You

You don't need the most expensive drone on the market to get started. You need the right drone for the job—and for most Midwest row crop farmers, that's a lot simpler (and cheaper) than the salespeople want you to believe.

Walk into any drone dealer or browse any agricultural technology website, and you'll quickly find yourself overwhelmed. Multispectral sensors. RTK positioning systems. Thermal cameras. Spray drones. Software subscriptions that cost more per month than your cell phone bill. The message is clear: precision agriculture requires serious investment.

Here's what they don't tell you: for basic crop monitoring—the kind that can genuinely help you make better decisions about your fields—you can get started for around $1,500. Sometimes less. And that entry-level setup will handle 80% of the useful applications most farmers actually need.

This chapter will help you separate the essential from the expensive, the practical from the promotional. We'll cover exactly what equipment you need to start, what you can safely skip, and how to think about upgrades down the road. The goal is simple: to get you flying and collecting useful data without emptying your operating account.

The Minimum Viable Setup

Let's start with what you actually need to fly over your fields and capture useful imagery. The list is shorter than you might expect.

The Drone Itself

For agricultural scouting and basic crop monitoring, the sweet spot in early 2026 is an entry-level consumer drone with a good RGB (standard color) camera. The leading option is the DJI Mini 4 Pro or its successor, the DJI Mini 5 Pro, which launched in September 2025.

Why these particular drones? Several reasons:

- Weight under 250 grams. This is the magic number. Drones weighing less than 0.55 pounds (250 grams) and flown recreationally don't require FAA registration. Even if you're flying commercially under Part 107, the lightweight design makes them easier to transport and less intimidating to learn on.
- Excellent camera quality. The Mini 4 Pro features a 1/1.3-inch sensor with 48MP photos and 4K/60fps video. The Mini 5 Pro upgrades to a full 1-inch sensor with 50MP resolution—the same sensor class used in much more expensive professional drones.
- Obstacle avoidance. Both models feature omnidirectional obstacle sensing – cameras and sensors that detect objects in all directions. The Mini 5 Pro adds forward-facing LiDAR for better performance in low-light conditions. This is crucial for a beginner.
- Reasonable flight time. The Mini 4 Pro offers 34 minutes per battery (45 with Plus battery). The Mini 5 Pro extends to 36 minutes with the standard battery, or up to 52 minutes with the Plus battery. In real-world agricultural use, expect 25–40 minutes per battery – enough to cover 40–100 acres depending on flight altitude and wind.
- Proven reliability. DJI dominates the consumer and prosumer drone market for good reason. Their flight controllers are stable, their apps are polished, and when something goes wrong at 200 feet, you want equipment tested by millions of users.

CURRENT PRICING (FEBRUARY 2026)

DJI Mini 4 Pro (RC-N2): $759

DJI Mini 4 Pro Fly More Combo (RC 2): $1,099 (3 batteries, *charging hub, bag)*

DJI Mini 5 Pro (RC-N3): $749–899

DJI Mini 5 Pro Fly More Combo: $1,049–1,279

Note: As of February 2026, DJI products face U.S. import restrictions. Check availability from authorized resellers.

The "Fly More" Question

Should you buy the basic drone or the "Fly More Combo"? For agricultural use, the Fly More Combo is almost always worth the extra cost. The combo includes three batteries instead of one – when flying fields, one battery isn't enough. You'll land after 25 minutes, swap batteries, and want to keep going.

The combo also includes a charging hub that charges multiple batteries in sequence – plug them all in and walk away. For a working farmer with limited patience for gadgets, the charging hub is worth its weight in gold.

What You DON'T Need to Start

Now let's talk about the equipment that drone dealers will try to sell you – and why you should politely decline, at least for now.

Multispectral Sensors ($5,000–$6,000+)

Multispectral cameras capture light beyond what the human eye can see, particularly in the near-infrared and red-edge wavelengths that reveal plant stress before it's visible. They're the foundation of "true" NDVI and other vegetation indices.

They're also expensive. The DJI Mavic 3 Multispectral runs approximately $4,959 and up – more than three times the cost of a capable RGB drone.

Here's the thing: a standard RGB camera can accomplish most of what you need for basic crop monitoring. Stand counts, emergence assessment, drainage problems, weed patches, obvious stress patterns, damage assessment, tile line mapping – all visible in regular color imagery.

Multispectral becomes valuable when you're ready to create variable-rate nitrogen prescriptions or detect stress 1–2 weeks before it's visible. If you're not set up for variable-rate application—if you don't have equipment to act on that data—the investment is premature.

RTK/PPK Positioning Systems ($2,000–$4,000+)

RTK (Real-Time Kinematic) and PPK (Post-Processed Kinematic) systems provide centimeter-level GPS accuracy. Essential for survey-grade mapping, but for basic crop scouting? Massive overkill. Standard GPS is accurate to within a few meters – enough to identify which part of the field has a problem.

Expensive Software Subscriptions ($100–$300/month)

Companies like DroneDeploy and Pix4D offer sophisticated cloud-based processing. At $200/month, you're spending $2,400 per year on software – more than the cost of your drone.

For basic monitoring, free or low-cost alternatives work fine:

- DJI Fly app (free): Comes with your drone. Basic mapping and flight planning built in.
- OpenDroneMap (free): Open-source processing software. Steeper learning curve, but capable.
- Pix4Dfields ($1,650/year, or $3,990 perpetual license): Agriculture-specific, best-in-class for multispectral processing. Spray Drones ($15,000–$30,000+) For Midwest row crop operations, the economics rarely work out for drone ownership. Limited capacity (2–10 gallons), regulatory complexity (Part 137 certification), and slow coverage (15–40 acres/hour vs. 200+ for ground sprayers). If you need aerial application, hire a service at $8–15 per acre. The Complete Starter Kit Based on the recommendations above, here's what a complete entry-level crop monitoring kit looks like:

ITEM PRICE NOTES

- DJI Mini 4/5 Pro Fly $1,049-$1,279
- Extra battery (fourth) $65-$100 (Optional)
- MicroSD card (128GB, $20-$40 Fast card U3/V30)
- Landing pad $15-$30 (Keeps dust out)

- ND filter set + sun $40-$70

TOTAL INVESTMENT	$1,189–$1,519	(estimated)

Call it $1,500 for a nicely equipped starter kit. That's less than most farmers spend on a single load of seed corn. And unlike the seed, the drone is reusable year after year.

Hidden Costs to Consider

FAA Part 107 Certification

If flying commercially – using drone data for farm management decisions – you technically need FAA Part 107 certification. The test costs $175, valid for two years. It covers airspace, weather, regulations, and basic aeronautical knowledge. Most people pass with a week or two of study.

Insurance

Drone liability insurance typically runs $500–$1,200 per year. Check with your farm insurance agent—many policies either cover drone operations or can add an endorsement for minimal cost.

Battery Replacement

Lithium-polymer batteries degrade over time. Expect to replace every 200–400 charge cycles, which is estimated to be roughly every 2–3 years. Budget $150–$300 for replacement batteries down the road.

The Upgrade Path: When to Invest More

Upgrade When...

- You've flown 50+ hours and feel limited by your current equipment.
- You're actively using drone data to make management decisions.
- You have variable-rate equipment and want to create your own prescriptions.
- You've identified a specific capability gap that better equipment would solve. Don't Upgrade When...
- You're buying equipment hoping to find a use for it.
- You haven't mastered your current equipment yet.
- A salesperson told you that you "need" something more sophisticated.

KEY PRINCIPLE

Don't buy equipment hoping to find a use for it. Buy equipment that solves a problem you've already identified. Prove the

concept with simple tools before scaling the investment.

Chapter Summary

- You can start effective crop monitoring for approximately
- $1,200–$1,500—a complete kit with drone, extra batteries, and essential accessories.
- Entry-level drones like the DJI Mini 4 Pro and Mini 5 Pro offer excellent capability – under 250g weight, 34–52 minute flight times (depending on battery), high-quality cameras, and obstacle avoidance.
- Skip the expensive add-ons for now: multispectral sensors, RTK positioning, pricey software subscriptions, and spray drones.
- A standard RGB camera can accomplish most of what you need: stand counts, drainage mapping, stress identification, and damage assessment.

- Upgrade only when you've identified a specific need – not because a salesperson told you to.

Chapter 2 Notes and References

1. DJI Store. "DJI Mini 4 Pro Specifications." Accessed February 2026. https://store.dji.com/product/dji-mini-4-pro

2. DJI. "DJI Mini 5 Pro." Product launch September 2025. https://www.dji.com/mini-5-pro

3. Federal Aviation Administration. "How to Register Your Drone." https://www.FAA.gov/uas/getting_started/register_drone

4. Advexure. "DJI Mavic 3 Multispectral: Top Precision Agriculture Drone in 2025?" February 2025.

5. Tom's Guide. "DJI Mini 5 Pro Review." October 2025. https://www.tomsguide.com/cameras-photography/drones/dji-mini-5-pro-review

Note: Equipment specifications and pricing accurate as of February 2026. Verify current availability, particularly given evolving U.S. import regulations affecting DJI products.

A NOTE ON TIMING: THE FCC COVERED LIST (DECEMBER 2025)

On December 22, 2025, the FCC issued Public Notice DA-25-1086, placing all foreign-manufactured UAS – including all DJI and Autel models – on the FCC Covered List.

What this means for buyers:

DJI drones already FCC-authorized remain legal to purchase and fly. Existing equipment authorizations are not revoked.

New DJI models cannot receive FCC authorization going forward. This affects future product launches, not current inventory. The $1,200–$1,500 entry point in this chapter remains valid for current inventory. The DJI Mini 4 Pro and Mini 5 Pro are available through authorized retailers as of early 2026. No U.S.-manufactured drone currently matches DJI at this price point. See Chapter 3 for a full discussion of alternatives.

(Monitor developments at suacoalition.org or auvsi.org)

Chapter 3: The Legal Stuff – FAA Rules That Actually Affect You

You don't need a law degree to fly a drone over your cornfield. But you do need to know the handful of rules that especially matter—and more importantly, which ones you can stop worrying about.

If there's one topic guaranteed to make a farmer's eyes glaze over, it's federal regulations. You've already got enough paperwork between FSA offices, crop insurance forms, and whatever new compliance requirement showed up in the mail this week. The last thing you want is another set of rules to learn before you can use a piece of equipment you already paid for.

Here's the good news: the FAA rules that apply to agricultural drone use are simpler than most people think. If you're flying a small drone over your own fields in rural Ohio or Indiana, you're operating in just about the most regulation-friendly environment possible. No restricted airspace. No crowds to avoid. No buildings to navigate around. Most of the FAA's drone regulations were written with urban areas and commercial operations in mind—not a farmer checking his soybeans from 200 feet up.

This chapter strips away the legal jargon and gives you exactly what you need to know. We'll cover when you need certification and when you don't, the basic flight rules that apply to agricultural use, how to check your airspace in about thirty seconds, and the simple record-keeping that protects you if questions ever arise. We'll also be honest about the gray areas—because there are a few.

Recreational vs. Commercial: Which One Are You?

This is the first question you need to answer, and for most farmers, it's more straightforward than the internet makes it sound. The FAA divides drone operations into two main categories, and which one you fall into determines what rules apply.

Recreational Flying

If you're flying a drone over your own fields to look at your own crops and you're not getting paid by anyone else to do it, you're likely a recreational flyer. The FAA calls this flying for "purely recreational purposes."

As a recreational flyer under the updated rules established by the FAA Reauthorization Act of 2024, you need to follow a short list of requirements. You must register your drone if it weighs more than 0.55 pounds (250 grams)—though if you took our advice in Chapter 2 and bought a DJI Mini 4 Pro or Mini 5 Pro, you're under that weight limit. You need to fly within visual line of sight, meaning you can see the drone with your own eyes at all times. You must stay at or below 400 feet above ground level, though for crop monitoring you'll rarely go above 300. You should avoid flying over people who aren't participating in the operation. And you need to stay clear of other aircraft and give way to manned aircraft if you encounter them.

For a farmer standing in a hayfield launching a drone to look at his corn across the road, these rules barely register. You can see your drone. You're well under 400 feet. There's nobody around. The nearest manned aircraft is probably a crop duster operating three counties over!

Commercial Flying

Here's where it gets slightly more nuanced. If you're using drone imagery to make management decisions on someone else's farm—or if you're offering drone services for hire—that's commercial operation, and it requires an FAA Remote Pilot Certificate, commonly known as "Part 107."

But what about using it on your own farm? The FAA has acknowledged that a farmer flying over his own fields to check his own crops is engaging in recreational activity, even if the information gathered informs business decisions. The key distinction is whether you're being compensated by a third party for the flight itself. Checking your own nitrogen stress to decide whether to side-dress more fertilizer? That's recreational. Flying your neighbor's fields and charging him for the maps? That's commercial.

THE GRAY AREA

What if your neighbor asks you to fly his field "as a favor" and doesn't pay you? Technically still recreational. What if he hands you fifty bucks for gas? Now it's getting murky. What if you're a farm manager making decisions for an absentee landowner? The FAA hasn't issued specific guidance on every scenario.

The practical advice:

if you're only flying your own operation,

don't lose sleep over it. If there's any chance you'll fly for

others – even informally – get your Part 107. It's not that hard,

and it removes all ambiguity.

Part 107 Certification: What It Is and Whether You Need It

The FAA Remote Pilot Certificate under 14 CFR Part 107 is the license that authorizes commercial drone operations in the United States. If you decide to get it—and I'd recommend it for anyone who's serious about using drones for farm management—here's what's involved:

The Test

The Part 107 exam is a 60-question, multiple-choice knowledge test administered at FAA-approved testing centers. You'll find testing locations in most mid-sized cities—Dayton, Columbus, Indianapolis, and Fort Wayne all have centers. The test costs $175 and takes about two hours. You need a 70% to pass.

The exam covers five main topic areas:

Airspace classification and operating requirements. This is the most heavily tested area. You'll need to understand the different classes of airspace (B, C, D, E, and G) and what rules apply in each. For rural agricultural flying, you're almost always in Class G – uncontrolled airspace – which has the fewest restrictions.

Weather and micrometeorology. Basic weather theory, how to read METARs and TAFs (aviation weather reports), and how weather affects small aircraft. Farmers already understand weather better than most test-takers – you know what a developing thunderstorm looks like and when wind makes outdoor work impractical.

Drone loading and performance. How weight, altitude, temperature, and humidity affect drone performance. Straightforward material.

Operations and regulations. The specific rules governing commercial drone flight – altitude limits, visual line of sight requirements, night operations, operations over people. This is the material you'll use in practice.

Crew resource management and decision-making. Safe operating practices, risk assessment, and emergency procedures. Common-sense material for anyone accustomed to operating heavy equipment.

How to Prepare

Most people can pass the Part 107 exam with one to two weeks of focused study. You don't need flight experience or an aviation background. The material is straightforward—it's a knowledge test, not a skills test. Nobody's asking you to fly anything during the exam.

Several study options are available:

Free resources. The FAA's own study guide ("Remote Pilot – Small Unmanned Aircraft Systems Study Guide," FAA-G-8082-22) is available as a free PDF download. It covers everything on the test. It's dry reading, but it's thorough and it's free.

Online courses. Companies like Pilot Institute, DARTdrones, and the Drone Pilot Ground School offer structured courses ranging from free to around $150–$300. These break the material into video lessons with practice quizzes. If you learn better from video than from reading, they're worth considering.

Practice tests. Multiple free practice exams are available online. Take several before scheduling your real test. If you're consistently scoring 85% or higher on practice tests, you're ready.

STUDY TIP FOR FARMERS

The sectional chart questions trip up a lot of people, but here's the thing – you probably already know how to read a map better than most test-takers. Sectional charts are just aviation maps with their own symbols. Spend extra time learning the airspace symbols and you'll be fine. The weather questions will feel familiar if you've ever watched a forecast and decided whether to cut hay.

The Certification Timeline

Step	Details	Time / Cost
Create IACRA account	FAA's online registration system at www.iacra.faa.gov	30 minutes / Free
Study for exam	Use free FAA guide or paid online course	~$300
Schedule test	Find testing center at www.faa.psi.com	~$175
Pass exam	60 questions, need 70%	2 hours
Apply for certificate	Submit through IACRA	Free after passing
Receive certificate	Temporary certificate immediately; permanent by mail in 6–10 weeks	—

Your Part 107 certificate is valid for two years. Renewal requires completing an online recurrent training course through the FAA's WINGS system—it takes about an hour and is free. No retesting required.

Do I Need Part 107? A Practical Decision Guide

Let's cut through the ambiguity with a simple framework. Answer these questions honestly:

FLOWCHART: DO I NEED PART 107?

1. Will you ONLY fly over your own property to monitor your own crops?

- -> If yes, and no one is paying you to fly: Recreational rules apply. Part 107 not required.

2. Will you ever fly over someone else's land for their benefit?

- -> If yes: You need Part 107.

3. Will anyone ever compensate you (cash, trade, or otherwise) for flying?

- -> If yes: You need Part 107.

4. Will you use drone data in reports for landlords, lenders, or insurance?

- -> Gray area. Probably recreational if it's your operation, but Part 107 removes any doubt.

My recommendation: Get Part 107 anyway. $175 and a week of study buys you complete legal clarity, the ability to help neighbors, and the option to offer services down the road if you choose to. Think of it as cheap insurance.

The Rules That Actually Matter

Whether you're flying recreationally or under Part 107, a handful of core rules apply. Here they are in plain language, with the agricultural context that makes them practical.

Stay Under 400 Feet AGL

AGL means "above ground level"—400 feet above the terrain directly below you, not 400 feet above sea level. For crop monitoring, this rule is essentially a non-issue. Your optimal flight altitude for most agricultural imaging is 200–300 feet. At that height, you get excellent resolution while covering enough ground to be efficient. There's no reason to go higher for the applications described in this book.

To give you a sense of scale: a 20-story building is roughly 200 feet tall. At 250 feet, individual corn plants are clearly distinguishable in your camera. At 300 feet, you can easily see row patterns and color differences. You're well under the ceiling with plenty of room to spare.

Maintain Visual Line of Sight (VLOS)

You must be able to see your drone with your unaided eyes (glasses are fine, binoculars are not) at all times during flight. This doesn't mean you need to stare at the drone every second—you'll be glancing at your controller screen to check the camera feed and flight data. But you need to be able to look up and locate the drone whenever necessary.

In practical terms, a Mini-series drone is visible to most people out to about 1,500–2,000 feet in good conditions. Farther than that, and it becomes a speck. This means you may need to reposition yourself during a flight over large fields. For a 160-acre quarter-section (roughly 2,640 feet on a side), you might set up near the center or fly the field in two passes from different positions.

PRACTICAL TIP

On a bright, overcast day, small drones can be hard to spot against a white sky. Orientation lights (standard on DJI drones) help, but the real solution is planning your flight path so the drone stays at a comfortable visual distance. Most automated flight apps handle this well – set up your flight plan, launch from a central location, and the drone will work a systematic pattern within view.

Don't Fly Over People

The FAA prohibits flying over people who are not directly participating in the drone operation. For a farmer flying over empty fields, this rule barely applies. Your crops don't count as people. Your cows don't either, for the record.

Where it matters: if you're flying near a road and workers are in the field below, or if family members or neighbors are present, be aware of where people are. Part 107 allows flight over people with certain categories of drones—the DJI Mini series qualifies for the least restrictive Category 1 due to its weight under 250 grams. But common sense applies: don't hover directly over anyone's head.

Give Way to Manned Aircraft

If you see or hear a manned aircraft—a crop duster, a helicopter, a low-flying plane—you must yield the right of way. In practice, this means land your drone or bring it down to a safe altitude until the aircraft passes. At 200–300 feet in rural airspace, encounters with manned aircraft are uncommon but not impossible, especially during application season when crop dusters are active.

If you know a crop duster is working in your area, it's worth calling the operator beforehand. Most agricultural pilots appreciate the heads-up, and many ag aviation companies are happy to coordinate. A two-minute phone call avoids any potential conflict.

Night Operations

Part 107 pilots can fly at night provided the drone has anti-collision lighting visible for at least three statute miles. Recreational flyers can also fly at night under similar lighting requirements as of the updated 2024 rules. For crop monitoring, night flying has limited utility—your camera needs light to capture useful imagery. Dawn and dusk flights during long summer days, however, are perfectly practical and often offer excellent conditions with lower wind and softer light.

Weather Minimums

Under Part 107, you must have at least three miles of visibility and stay at least 500 feet below and 2,000 feet horizontally from clouds. For recreational flyers, the rules require flying in conditions that allow you to maintain visual line of sight. Either way, this is common sense you already practice—you wouldn't launch a drone into fog or a thunderstorm any more than you'd start combining in a downpour.

Checking Your Airspace: The 30-Second Task

One of the most common concerns new drone pilots have is airspace—"Am I allowed to fly here?" For most Midwest farmers, the answer is almost certainly yes. But it takes thirty seconds to verify, so let's walk through it.

Understanding Airspace Classes

The FAA divides the national airspace system into classes, each with different rules. Here's the short version:

Airspace Class	Where You'll Find It	Can You Fly?
Class G (Uncontrolled)	Covers most rural areas away from airports	No authorization needed
Class E (Controlled)	Covers some rural areas, particularly around smaller airports and their surface boundaries	Authorization usually required; check airspace before flying
Class D	Surrounds airports with control towers, typically small to medium-sized facilities	LAANC authorization required

Airspace Class	Where You'll Find It	Can You Fly?
Class C	Surrounds mid-size airports such as Dayton and Columbus	LAANC authorization required
Class B	Surrounds major airports such as Chicago, Detroit, and Atlanta	LAANC authorization required; rarely granted for drone operations
Restricted / Prohibited	Military bases and other sensitive areas	Flight not permitted regardless of authorization

If your farm is more than five miles from an airport with a control tower, you're almost certainly in Class G airspace—no authorization needed, no hoops to jump through. That describes most of the agricultural land in Western Ohio and Eastern Indiana.

The Apps That Make It Easy

Two free apps will tell you everything you need to know about your airspace in seconds:

B4UFLY is the FAA's official app, available for iOS and Android. Open it, and it shows a map with your location and any airspace restrictions nearby. Green means "clear". Yellow means "check the details". Red means "don't fly without authorization". It's as simple as opening a weather app.

Aloft (formerly Kittyhawk) provides similar airspace information with the added ability to submit LAANC (Low Altitude Authorization and Notification Capability) requests directly through the app. If you happen to be near controlled airspace, LAANC authorization is usually processed automatically within seconds. It's a free, instant process, not the bureaucratic nightmare you might expect.

Both apps also show temporary flight restrictions (TFRs) – temporary airspace closures for things like presidential visits, major sporting events, or firefighting operations. In rural areas, TFRs are rare. But check before every flight. It takes ten seconds.

REAL-WORLD SCENARIO

Say your farm is 12 miles east of Dayton's James M. Cox International Airport. Open B4UFLY. Your location shows green – Class G airspace, no restrictions. You're good to go. It takes less time than checking the weather on your phone.

Now say you're farming near a small municipal airport with a control tower. Open Aloft, request LAANC authorization for the altitude you need (say 200 feet). Authorization comes back in 15 seconds. You're approved and logged.

Go and fly your fields.

Drone Registration: Quick and Simple

The FAA requires registration for any drone weighing more than 0.55 pounds (250 grams). If you bought a DJI Mini 4 Pro or Mini 5 Pro, you're right at or just under the threshold—these drones are designed specifically to come in under 250 grams. Without accessories, they don't require registration for recreational use.

However, if you're flying under Part 107 (commercial), registration is required regardless of weight. And if you add accessories that push the weight over 250 grams, registration applies.

The registration process is straightforward:

Go to www.faadronezone.faa.gov. Create an account, enter your drone information, pay the $5 registration fee. You'll receive a registration number that you mark on your drone (a small label is fine). Registration is valid for three years.

That's it. Five dollars. Five minutes. Three years of coverage. And you can register multiple drones under the same registration number if you're a Part 107 pilot.

Remote ID: The New Requirement

Starting in March 2024, the FAA requires most drones to broadcast Remote ID—a digital "license plate" that transmits the drone's identity, location, altitude, and control station location. This allows the FAA and law enforcement to identify drones in flight.

The good news for most buyers: any drone manufactured after September 2022 has Remote ID built in. The DJI Mini 4 Pro, Mini 5 Pro, and virtually every current consumer drone complies automatically. You don't need to install anything extra or configure anything special – it's active by default.

If you're flying an older drone manufactured before the deadline, you have two options: add an aftermarket Remote ID broadcast module (typically $100–$200) or fly only at FAA-Recognized Identification Areas (FRIAs). For most farmers buying new equipment, Remote ID is a non-issue—it's already handled.

Record-Keeping: Protecting Yourself

The FAA doesn't require recreational pilots to keep formal flight logs. Part 107 pilots aren't explicitly required either, though it's strongly recommended. Regardless of which category you're in, I'd recommend keeping simple records. Here's why:

Regulatory protection. If anyone ever questions your drone operations, a flight log demonstrates that you were following the rules – flying at legal altitudes, in appropriate conditions, within approved airspace.

Insurance documentation. If you need to make an insurance claim – either for drone damage or for crop damage you're documenting – flight records support your case.

Operational improvement. Tracking your flights helps you identify patterns. Which flights produced the best data? What conditions gave you problems? Over a season, your log becomes a reference guide for planning future flights.

Your flight log doesn't need to be complicated. A simple notebook or spreadsheet works fine. Record the date, time, location (field name or number), weather conditions, flight altitude, battery used, and any notes about what you observed or any problems encountered.

SAMPLE FLIGHT LOG ENTRY

Date: *June 15, 2026*

Time: *10:30 AM – 11:05 AM*

Location: *Field 3 (North 80)*

Weather: *Partly cloudy, 78°F, wind SW 8 mph*

Altitude: *250 ft AGL*

Equipment: *DJI Mini 4 Pro, Battery #2*

Purpose: *V8 nitrogen assessment, NDVI baseline*

Notes: *Yellow streaking visible NE quarter. Ground-truth needed.*

Accomplished: *127 images captured.*

A flight log template is included in Appendix H. You can photocopy it, keep it in the truck, and fill it out in two minutes after each flight. Some pilots prefer digital logs – the apps mentioned above (Aloft, DJI Fly) also automatically record flight data that you can export later.

What You Don't Need to Worry About

Let's clear up some common misconceptions that keep people from getting started.

"Do I need permission to fly over my own land?"

No. You own the land; the airspace above it (up to a certain altitude) is available for drone operations just like it's available to any aircraft. You don't need to notify anyone, file a flight plan, or get permission from your township, your county, or your congressman. The FAA has exclusive authority over the national airspace, and they've already set the rules you need to follow.

"What about flying over my neighbor's land?"

The FAA regulates the airspace, not the land underneath it. Legally, you can fly a drone over your neighbor's property—airspace isn't owned by the landowner. That said, common courtesy applies. If your flight path crosses a neighbor's property, a quick heads-up goes a long way in maintaining good relationships. Rural communities run on trust, and a five-minute conversation prevents misunderstandings.

"Will I get in trouble if something goes wrong?"

If you're following the rules—appropriate airspace, legal altitude, visual line of sight—and a gust of wind brings down your drone, that's an accident, not a violation. The FAA expects drone pilots to operate responsibly, not perfectly. If a crash causes property damage or injury, you may have a civil liability issue (which is what insurance is for), but you're not going to have the FAA knocking on your door because your drone clipped a power line during an unexpected wind gust.

That said, reckless or intentional violations are taken seriously. Don't fly near airports without authorization. Don't fly over emergency scenes. Don't interfere with manned aircraft. Use common sense, and you'll be fine.

"Do I need a special license to fly near my grain bins or buildings?"

No. The FAA's 400-foot altitude limit is measured from the ground, and there's an exception that allows you to fly above 400 feet if you're within 400 feet of a structure and don't exceed 400 feet above that structure. For practical agricultural use, this rarely matters—your grain bins aren't 400 feet tall. Fly at your planned altitude, and the structures on your property aren't an issue.

A Note on State and Local Laws

While the FAA has primary authority over airspace and drone operations, some states and localities have passed their own laws regarding drones. Ohio law, for example, generally follows federal guidelines and has not imposed significant additional restrictions on agricultural drone use. Indiana similarly defers largely to federal rules.

A few states have enacted privacy-related drone laws—restrictions on using drones for surveillance of private property, for example. These are primarily aimed at peeping drones in suburban neighborhoods, not farmers monitoring their fields. But it's worth being aware that the legal landscape can vary by state, and new legislation is always possible.

The practical takeaway: follow the FAA rules, be a good neighbor, use your drone for legitimate agricultural purposes, and you're on solid legal ground.

Drone Insurance: Worth Considering

Drone liability insurance isn't legally required for recreational or most Part 107 operations, but it's worth considering. There are two types:

Hull insurance covers damage to your drone. If a $1,500 drone meets an unexpected fence post, hull insurance pays for repair or replacement. Given the relatively low cost of entry-level drones, many farmers self-insure this risk.

Liability insurance covers damage your drone causes to other people or property. This is the most important coverage. If your drone somehow damages a power line, a vehicle, or injures someone, liability insurance protects you.

Several companies offer drone-specific coverage. Annual policies typically run $500–$1,200 for $1 million in liability coverage. Some companies offer pay-per-flight options for occasional users at $5–$10 per flight.

Before buying separate drone insurance, check with your existing farm insurance agent. Many farm liability policies either already cover drone operations as an incidental activity or can add coverage through an endorsement for a modest premium increase. Your agent can advise on the most cost-effective approach.

A Note on Evolving Regulations and DJI

It would be irresponsible to write about drone regulations in 2026 without addressing the elephant in the room: the regulatory uncertainty surrounding DJI, the Chinese drone manufacturer that dominates the consumer and agricultural drone market.

In late 2024, the U.S. House of Representatives passed the "Countering CCP Drones Act," which sought to add DJI to the FCC's covered list. On December 22, 2025, the FCC acted through Public Notice DA-25-1086, placing all foreign-manufactured UAS—including all current DJI and Autel models—on the FCC Covered List. This means new DJI drone models cannot receive FCC equipment authorization going forward. However, DJI products that were already FCC-authorized remain legal to purchase and fly in the U.S.; existing authorizations are not revoked.

However, the regulatory landscape is shifting. Several federal agencies have restricted or banned DJI drones for government use. Some state governments have followed suit. And the broader U.S.-China technology tensions continue to create uncertainty about long-term availability of DJI products.

What this means for you as a farmer:

Your existing DJI drone is not going to be "bricked" or made illegal. Even in a worst-case regulatory scenario, existing equipment would remain legal to operate.

DJI remains the best value for agricultural monitoring in 2026. No competitor offers the same combination of capability, reliability, and price. The recommendations in Chapter 2 still stand.

Alternatives exist and are improving. U.S.-manufactured drones from companies like Skydio are viable options, though generally at higher price points for comparable capability. Note that Autel is also Chinese-manufactured and subject to the same FCC Covered List restrictions as DJI. See Chapter 3 for a full breakdown of current alternatives.

Stay informed. Monitor developments through the Small UAV Coalition, the Association for Unmanned Vehicle Systems International (AUVSI), or simply check drone news periodically. Changes won't happen overnight.

The core advice of this book – start simple, prove the value, then invest more – is equipment-agnostic. Whatever brand of drone you fly, the regulations, the science, and the agronomic applications all work the same way.

Chapter Summary

- For most farmers flying over their own fields, recreational rules apply and Part 107 certification is not required—but getting it anyway ($175, one to two weeks of study) provides complete legal clarity and future flexibility.
- The rules that matter are simple: stay under 400 feet, keep the drone in sight, don't fly over people, yield to manned aircraft, and check your airspace before flying.
- Most rural agricultural land sits in Class G (uncontrolled) airspace with zero restrictions. Free apps like B4UFLY and Aloft let you verify in seconds.
- Drone registration costs $5 for three years. Drones under 250 grams
- (like the DJI Mini series) may not require registration for recreational use.
- Remote ID is built into any drone manufactured after September
- 2022 – no extra hardware or configuration needed.
- Keep a simple flight log for every flight. It protects you legally, supports insurance claims, and improves your operations over time.
- Drone insurance is worth considering – check with your farm insurance agent first, as coverage may already be included or available as a low-cost endorsement.

- DJI products that were FCC-authorized prior to December 22, 2025, remain legal to purchase and fly. New DJI models cannot receive FCC authorization under Public Notice DA-25-1086. No U.S.-made entry-level alternative currently matches DJI's price and capability—see Chapter 3 for alternatives.

Chapter 3 Notes and References

1. Federal Aviation Administration. "Recreational Flyers & Modeler Community-Based Organizations." Updated 2024. https://www.faa.gov/uas/recreational_flyers

2. Federal Aviation Administration. "14 CFR Part 107 – Small Unmanned Aircraft Systems." https://www.ecfr.gov/current/title-14/chapter-I/subchapter-F/part-107

3. Federal Aviation Administration. "Become a Drone Pilot: Remote Pilot Certificate." https://www.faa.gov/uas/commercial_operators/become_a_drone_pilot

4. Federal Aviation Administration. "Remote Pilot – Small Unmanned Aircraft Systems Study Guide." FAA-G-8082-22. https://www.faa.gov/regulations_policies/handbooks_manuals/aviation/remote_pilot_study_guide

5. Federal Aviation Administration. "UAS Data Exchange (LAANC)." https://www.faa.gov/uas/programs_partnerships/data_exchange

6. Federal Aviation Administration. "Remote Identification of Unmanned Aircraft." Final Rule, effective March 16, 2024. https://www.faa.gov/uas/getting_started/remote_id

7. Federal Aviation Administration. "FAA DroneZone." https://faadronezone.faa.gov

8. Federal Aviation Administration. "Airspace 101 – Rules of the Sky." https://www.faa.gov/uas/getting_started/airspace

9. U.S. Congress. "Countering CCP Drones Act." H.R. 2864, 118th Congress. https://www.congress.gov/bill/118th-congress/house-bill/2864

10. Ohio Revised Code, Chapter 4561: Aeronautics. Provisions related to unmanned aircraft systems.

Note: FAA regulations are subject to change. Verify current rules at www.faa.gov/uas before flying. This chapter provides general guidance and does not constitute legal advice.

PART TWO — The Science Made Simple – How Plants Talk Through Light

Chapters 4–7

Drone maps are only useful if you understand what they're showing you. Part II explains the science behind agricultural remote sensing in language built for farmers, not physicists. Chapter 4 covers how light interacts with crops—why healthy corn reflects more near-infrared light, and why that matters. Chapter 5 introduces the vegetation indices (NDVI, NDRE, and others) that turn raw sensor data into color-coded maps you can act on. Chapter 6 contrasts the RGB and multispectral cameras you'll choose between—what each one sees, and what each one misses. Chapter 7 turns that comparison into a financial decision, with break-even calculations by acreage that show when multispectral pays for itself and when RGB is the smarter choice. These four chapters give you the foundation to read every map in the rest of the book.

Chapter 4: Light, Color, and What Your Crops Are Telling You

Every plant in your field is broadcasting information about its health, all day, every day. You just can't see most of it. This chapter explains what your crops are saying—and how a camera can translate.

Before we talk about flight plans and software, we need to spend a chapter on the science that makes all of this work. Don't worry—this isn't a college lecture. There are no equations to memorize and no lab reports to turn in. But understanding the basic principles of how plants interact with light is the difference between staring at a pretty map and actually knowing what it means.

The good news is that the core concept is simple enough to explain over a cup of coffee. The even better news is that once you understand it, everything else in this book—vegetation indices, multispectral sensors, crop health maps—will make intuitive sense. You'll know why the maps look the way they do, what the colors represent, and when to trust the data versus when to be skeptical.

So let's start with something you already know: why plants are green.

Why Plants Are Green (And Why That Matters)

You've looked at green plants your entire life, but have you ever stopped to think about why they're green? It's not decorative. It's chemistry.

Leaves are full of a pigment called chlorophyll. Chlorophyll is the engine of photosynthesis—it's what allows a plant to capture energy from sunlight and turn it into sugars for growth. But chlorophyll doesn't use all of the light that hits a leaf. It's picky.

Chlorophyll absorbs red light and blue light very efficiently. These are the wavelengths it needs to power photosynthesis. But green light?

Chlorophyll mostly bounces green light right back. It doesn't need it. That reflected green light is what reaches your eyes, and it's why a healthy cornfield looks green in July.

Here's the key insight: the color you see when you look at a plant is the light the plant didn't use. It's the leftovers. A healthy plant with lots of chlorophyll absorbs most of the red and blue light hitting it and reflects back the green. That's the signal your eyes pick up.

THE SIMPLE VERSION

Sunlight hits a leaf. The leaf absorbs red and blue light for energy (photosynthesis). The leaf reflects green light back because it doesn't need it. Your eyes see the reflected green light. That's why plants look green. The color you see is the light the plant rejected.

Now, this should already tell you something useful. If a plant is green, it means it has plenty of chlorophyll and it's actively photosynthesizing. If a plant starts turning yellow, it means chlorophyll levels are dropping. The leaf is losing its ability to absorb red and blue light, so more of those wavelengths bounce back along with the green – and the mix shifts toward yellow.

Every farmer knows this intuitively. A field that's deep, dark green is a healthy field. A field that's turning pale or yellow is telling you something is wrong – maybe nitrogen deficiency, maybe disease, maybe waterlogging. You've been reading this signal your entire farming career. You just didn't think of it as "remote sensing."

But here's the problem with relying on what you can see: by the time a plant looks yellow to your eyes, the stress is already well advanced. The chlorophyll loss that causes visible yellowing represents a significant decline in photosynthetic capacity. Yield potential has already been lost. The plant has been struggling for days or weeks before it started showing visible symptoms.

Which brings us to the most important part of this chapter: the light you can't see.

The Secret Color You Can't See

The light your eyes can detect—the rainbow from violet to red—is called visible light. It's a tiny sliver of all the electromagnetic energy bouncing around us. Just beyond the red end of the visible spectrum sits a type of light called near-infrared, or NIR. You can't see it, but it's there, and it's the key to everything that follows in this book.

Here's what makes NIR so important for agriculture: healthy plants reflect near-infrared light like a mirror.

This isn't a subtle effect. A healthy leaf reflects somewhere around 10–15% of the visible green light that hits it. But that same leaf reflects 40–60% of the near-infrared light. Healthy plants practically glow in the near-infrared. If your eyes could see NIR, a July cornfield would be blindingly bright—far brighter than it appears in visible light.

Why? It comes down to leaf structure. Inside a healthy leaf, the cells are tightly organized and full of water. This internal structure acts like a highly efficient reflector for near-infrared wavelengths. The light enters the leaf, bounces around inside the cell structure, and a large percentage of it gets reflected back out. The plant doesn't use NIR for photosynthesis – so it sends it right back.

Now here's where it gets interesting—and useful.

When a plant becomes stressed, its internal cell structure starts to break down. The cells lose water. The organized structure that was so good at reflecting NIR starts to collapse. And the amount of near-infrared light the leaf reflects drops – sometimes dramatically.

This change in NIR reflectance happens before the plant shows any visible symptoms. The leaf still looks green to your eyes. The chlorophyll hasn't broken down enough for you to see yellowing. But inside the leaf, the cellular structure is already deteriorating, and the NIR reflectance is already declining. A camera sensitive to near-infrared light can detect this change days or even weeks before you'd notice anything wrong standing in the field.

KEY INSIGHT

A stressed plant changes how it reflects near-infrared light before it changes how it looks to your eyes. That's the entire foundation of drone-based crop monitoring. NIR cameras don't see the plant as you see it – they see what's happening inside the leaf.

The One Principle That Explains Everything

If you take nothing else from this chapter, take this:

THE FUNDAMENTAL PRINCIPLE

Healthy plant = absorbs red light (for photosynthesis) + reflects lots of NIR light (healthy cell structure)

Stressed plant = reflects more red light (less chlorophyll) + reflects less NIR light (damaged cell structure)

That's it. That's the whole foundation of remote sensing for crop health. Every vegetation index, every crop health map, every "precision agriculture" analysis comes back to this basic comparison: how much red light is a plant reflecting versus how much near-infrared?

When you see a crop health map with green and red zones, what you're really looking at is a visualization of this comparison across your entire field. The green zones are areas where plants are absorbing lots of red light and reflecting lots of NIR – the signature of health. The red or yellow zones are areas where more red light is bouncing back and less NIR is reflecting – the signature of stress.

The actual math behind the maps is straightforward—we'll cover specific vegetation indices in Chapter 5. But the concept is always the same: compare how much red light a plant reflects to how much NIR it reflects, and you get a measure of how healthy it is.

Why Timing Matters: Seeing Stress Before It Costs You

Here's a scenario every corn farmer knows. You're driving past a field in late June, and everything looks fine – a uniform canopy of dark green. Three weeks later, you're in that same field and you notice a patch in the northwest corner where the lower leaves are firing yellow. Nitrogen deficiency. The corn is at V12 or VT, and you've already lost yield potential in that area.

If you'd known about the problem at V8 – back when the field still looked uniformly green to your eyes – you could have made a side-dress application to that area. The cost of an extra pass with the applicator over a few acres would have been trivial compared to the yield loss you're now staring at.

This is the timing advantage of NIR-based monitoring. Research consistently shows that changes in near-infrared reflectance precede visible stress symptoms by one to three weeks, depending on the type and severity of stress. That window – between when a camera can detect a problem and when your eyes can – is the window where intervention is most effective and least expensive.

Type of Stress	Visible Symptom Timing	NIR/Early Detection Timing
Nitrogen Deficiency	Visual symptoms appear as lower leaf firing between V10 and VT.	Drone sensors detect changes in cell structure as early as V6–V8 — roughly 2 to 3 weeks before the problem is visible from the ground.
Water Stress (Drought)	Visual symptoms appear as leaf rolling and wilting.	Drone sensors detect the stress before visible rolling begins — approximately 1 to 2 weeks earlier.

Type of Stress	Visible Symptom Timing	NIR/Early Detection Timing
Disease Onset (e.g., Tar Spot)	Visual symptoms appear as visible lesions on leaves.	Drone sensors detect reduced NIR reflectance 1 to 2 weeks before lesions spread to a visible level.
Herbicide Injury	Visual symptoms appear as leaf curling and discoloration.	Drone sensors detect cell structure damage several days before it becomes visible.
Root Damage (Compaction, SCN)	Visual symptoms appear as stunted growth and yellowing.	Drone sensors detect reduced vigor in NIR reflectance 1 to 3 weeks before visible symptoms emerge.

Not every stress event gives you the same early warning. Sudden catastrophic damage – a hailstorm, severe flooding – shows up in visible light immediately and doesn't need NIR to detect. The real value of near-infrared monitoring is with chronic, progressive stress: the kind that creeps up on you over days and weeks, doing incremental damage before you'd ever notice it from the road.

The Cattle Temperature Analogy

If the science of plant reflectance still feels abstract, try this analogy.

You're a cattleman checking on your herd. Every animal looks fine – they're eating, they're moving around, their coats look normal. But you pull out a thermometer and check temperatures. One heifer is running 103.5°F – a degree and a half above normal. She doesn't look sick yet. She's not off her feed yet. But something is brewing. You separate her, monitor closely, and catch the respiratory infection before it spreads to the rest of the herd.

That's exactly what near-infrared monitoring does for your crops. The field looks green and healthy – that's the equivalent of the cattle looking normal. But the NIR camera takes the field's "temperature" and finds areas that are slightly off. The numbers don't match what a perfectly healthy plant should be producing. Something is brewing. You ground-truth those areas, identify the problem, and intervene before it costs you yield.

You can't see a fever. You can't see declining NIR reflectance. But in both cases, you can measure it – and acting on what you measure, before symptoms become obvious, is the whole point.

What Your Regular Camera Can (and Can't) Tell You

At this point you might be thinking: "Great, so I need an expensive multispectral camera to see any of this." Not necessarily. Let's be clear about what the standard RGB camera on your drone—the one you already have—can and can't do.

What RGB Can See

A regular camera sees the same three colors your eyes see: red, green, and blue. It captures visible light. And visible light carries a surprising amount of useful information:

Color differences. If part of your field is a different shade of green – or is turning yellow or brown – your camera captures that. From 250 feet up, you see patterns that are invisible from ground level. Streaks, patches, gradients, and boundaries that the bird's-eye view reveals.

Structural differences. Missing plants, short plants, lodged plants – all visible in RGB imagery. Stand counts, emergence uniformity, hail damage, wind damage, wildlife damage. These don't require special sensors.

Water and drainage patterns. Wet spots, ponding, and drainage problems show up clearly in color imagery. After a rain event, an RGB drone flight can map your tile line performance better than any diagnostic you could do from the ground. You can literally see which tiles are draining and which aren't.

Weed pressure. Before canopy closure, weeds between rows are visible from above. Patches of herbicide-resistant waterhemp or marestail stand out against the pattern of crop rows. You can create spot-spray maps and avoid blanket-treating entire fields.

Equipment problems. Planter skips show up as gaps in the row pattern. Fertilizer spreader overlap or skip creates streaks parallel to the application direction. Spray drift from a neighbor's field creates edge effects. All visible in standard color imagery.

What RGB Can't See

The limitation of a standard camera is simple: it can't see near-infrared light. That means it can't detect the early stress signals we just discussed. By the time stress shows up in visible light – in the red, green, and blue wavelengths – the problem is already advanced enough that you could probably spot it from the ground if you happened to be standing in the right place.

Specifically, RGB cameras cannot reliably:

Detect stress before visible symptoms. This is the big one. The entire early-warning advantage of NIR is off the table with a standard camera.

Calculate accurate vegetation indices. Some software programs will generate an "NDVI" map from RGB imagery. This is a mathematical approximation—it uses the visible red channel as a substitute for the comparison between red and NIR. It can reveal patterns, but it's not true NDVI, and the values aren't calibrated for precise decision-making.

Assess nitrogen status with precision. Nitrogen deficiency eventually shows as yellowing, which RGB captures. But by that point, you're past the optimal intervention window. True nitrogen assessment requires NDRE or similar indices based on near-infrared and red-edge wavelengths.

THE BOTTOM LINE ON RGB

An RGB camera is a powerful tool for scouting, stand assessment, drainage mapping, and identifying visible problems across your fields. It covers roughly 80% of the practical uses most farmers need. The remaining 20% – early stress detection, precise nitrogen mapping, calibrated vegetation indices – requires multispectral capability. Chapter 6 explores this trade-off in detail, and Chapter 7 helps you decide if and when to invest in multispectral.

A Farmer's Guide to the Light Spectrum

We've been talking about visible light and near-infrared as if they're two different things. They are – but they're also part of the same continuous spectrum. Understanding where the key wavelengths sit helps you make sense of the sensors and indices you'll encounter in later chapters.

Think of the electromagnetic spectrum as a ruler. On one end, you have short wavelengths – ultraviolet light, the kind that gives you a sunburn. In the middle, you have the wavelengths your eyes can see: violet, blue, green, yellow, orange, and red. And just past red – at wavelengths slightly longer than your eyes can detect – you enter the near-infrared zone.

For agricultural remote sensing, five regions of this spectrum matter most:

Band	Wavelength	What It Tells You About Crops
Blue	450–490 nm	Chlorophyll and carotenoid content; water depth in surface water
Green	520–570 nm	The light plants reflect most in the visible range; general vigor
Red	630–690 nm	Chlorophyll absorption peak; strong indicator of photosynthetic activity

Band	Wavelength	What It Tells You About Crops
Red Edge	700–740 nm	Transition zone between red absorption and NIR reflection; highly sensitive to chlorophyll content and nitrogen status
Near-Infrared (NIR)	760–900 nm	Internal leaf structure and water content; strongest indicator of plant health

The "red edge" deserves special attention. It's the narrow band between visible red and near-infrared where plant reflectance changes rapidly. In a healthy plant, reflectance jumps sharply from low (red absorption) to high (NIR reflection) across this narrow band. When a plant is stressed, that sharp jump becomes more gradual – the transition softens. This is why the red-edge band is so sensitive to chlorophyll content and nitrogen status. It picks up subtle changes that neither the red band nor the NIR band alone would reveal.

You don't need to memorize wavelength numbers. What matters is understanding that different parts of the spectrum carry different information about your crops, and that the most valuable information – the early stress signals – lives in wavelengths your eyes can't see.

What Affects the Signal: It's Not Just the Plant

If this were a perfect world, the light reflecting off your field would tell you exactly what every plant is doing at every moment. In reality, several factors besides plant health affect what a drone camera captures. Understanding these factors helps you avoid misinterpreting your maps.

Soil Background

Early in the season, before the canopy closes, your camera sees a mix of plant and bare soil. Different soil types reflect light differently – dark organic soils reflect less than light-colored sandy or clay soils. This soil signal contaminates vegetation indices, making sparse stands look less healthy than they actually are in dark soil areas, and healthier than they actually are in light soil areas.

This is why some vegetation indices (like SAVI, which we'll cover in Chapter 5) include a correction factor for soil brightness. It's also why early-season drone maps need careful interpretation. If your field has variable soil types—and in Western Ohio, most do—the patterns you see in a V4 map might partly reflect soil differences rather than crop health differences.

Sun Angle and Shadows

The angle of the sun affects how much light hits your canopy and how much reflects back toward the camera. Flights at different times of day will produce slightly different results. Morning flights cast shadows westward; afternoon flights cast them eastward. The ideal window is within two hours of solar noon on overcast days, when diffuse light minimizes shadow effects.

Partly cloudy days are the trickiest. A cloud shadow passing over part of your field during a flight creates a dark patch in the imagery that looks like crop stress but isn't. If your mapping software stitches together images taken seconds apart – some in sun, some in shadow – the resulting map can have misleading variability.

Crop Canopy Architecture

Not all crops present themselves the same way to an overhead camera. Corn, with its tall, upright leaves, reflects light differently than soybeans with their broad, horizontal leaf canopy. Even within a single crop, different hybrids and varieties may have slightly different canopy structures that affect reflectance. Row direction relative to sun position can also create differences.

This doesn't mean you can't compare across hybrids or crops – but it means you should be cautious about interpreting small differences between fields planted to different varieties. Within a single field planted to one hybrid, the variability you see is much more likely to reflect genuine differences in plant health.

Moisture on Leaves

Wet leaves reflect light differently than dry leaves. Dew, recent rain, or irrigation water on the canopy changes reflectance values, particularly in the near-infrared. For the most reliable data, fly after the canopy has dried – typically two or more hours after dew or rain.

PRACTICAL TAKEAWAY

The best conditions for reliable crop monitoring imagery: fly within two hours of solar noon, on overcast or consistently sunny days (not partly cloudy), after the canopy has dried, at a consistent altitude. You won't always get perfect conditions – Midwest weather doesn't cooperate on demand – but knowing what affects the data helps you interpret it with appropriate confidence.

"Aha Moment" Examples From Real Fields

Theory is useful. But the real understanding comes when you see this play out in your own fields. Here are common scenarios where the gap between what you see and what the camera sees becomes clear.

The Compaction Problem You Drove Past for Five Years

A farmer in Darke County, Ohio, had been noticing slightly lower yields in a section of his field near a grass waterway. From the road, the corn looked fine – maybe a touch shorter in that area, but nothing alarming. His first drone flight at V10 revealed a distinct circular pattern of reduced vigor centered on the spot where his grain cart made its turns during harvest. The compaction zone was nearly eight acres – far larger than he'd estimated from the ground. He deep-ripped the area that fall and picked up six bushels per acre the following year.

The Nitrogen Puzzle Solved From 250 Feet

A farm manager in Randolph County, Indiana, was using a uniform nitrogen rate across a 200-acre cornfield. Drone imagery at V8 showed a clear gradient of crop vigor running east to west across the field. The eastern third was noticeably more vigorous than the western third. From the ground, the difference was barely perceptible. Overlaying the drone map with a soil survey map revealed the answer: the soil transitioned from a Brookston silty clay loam in the east (higher organic matter, more mineralization) to a Miami silt loam in the west (lower organic matter). The eastern third was getting adequate nitrogen from soil mineralization plus the applied rate. The western third needed more. A variable-rate prescription the following year shifted 20 pounds of N from east to west. The result was a more uniform canopy and three additional bushels per acre field-wide.

The Tile Line Diagnosis After a June Rain

An Ohio farmer flew his fields two days after a three-inch rain in early June. The RGB imagery – no multispectral required – showed the drainage pattern as clearly as an X-ray. He could see every functioning tile line as a dark-green strip of healthy, well-drained crop. But one line, installed fifteen years earlier, showed a break: the dark-green strip ended abruptly, and a yellow-green wet area spread south of the interruption. He had the tile repaired that August. "I'd have never found that from the ground," he said. "The wet spot was in the middle of the field, and by the time it dried up after the rain, you couldn't tell it had been there."

The Planter Problem Hidden in Plain Sight

A V6 flight over a newly planted soybean field revealed distinct streaks running the direction of planting – alternating bands of slightly better and slightly worse emergence. The pattern was almost perfectly regular. From the ground, the field looked like a uniform stand. From 250 feet up, it was obvious: one unit on the planter was setting seed slightly deeper than the others, causing delayed emergence in those rows. The farmer recalibrated his planter before planting his remaining fields. "That drone paid for itself in one flight," he told a neighbor.

Connecting What You Know to What You Can See

The power of aerial monitoring isn't in the technology alone. It's in the combination of what the camera shows you and what you already know about your fields. Neither is complete without the other.

A drone map tells you where. Your experience tells you why. The camera can show you that a ten-acre patch in the south end of Field 5 is less vigorous than the rest. But it can't tell you whether that's because of nitrogen deficiency, compaction, drainage, a soil type transition, or the fact that the previous owner piled snow there for thirty winters. You know your fields. The aerial view fills in the gaps in your knowledge and gives you precise boundaries for the patterns you've always suspected were there.

This is why the "ground-truthing" step—walking the areas that show up as problems on your map—is non-negotiable. Maps tell you where to look. Your boots on the ground tell you what you're looking at. We'll cover systematic ground-truthing in Chapter 10, but the principle starts here: never spray, fertilize, or make a management change based solely on a map. Always confirm with your own eyes and your own knowledge first.

KEY PRINCIPLE

The technology shows you WHERE. Your experience tells you WHY. Neither is complete without the other. A drone map without ground-truthing is just a pretty picture. Ground-truthing without a map means you're only checking the places you already knew to look.

Chapter Summary

- Plants are green because chlorophyll absorbs red and blue light for photosynthesis and reflects green light back. The color you see is the light the plant didn't use.
- Healthy plants also reflect large amounts of near-infrared (NIR) light – a wavelength invisible to the human eye but detectable by cameras. This NIR reflectance is driven by the internal cell structure of the leaf.
- When a plant becomes stressed, its NIR reflectance drops before visible symptoms appear. This is the foundation of early stress detection—a one-to-three-week early warning window.
- The fundamental principle: healthy plants absorb red and reflect NIR; distressed plants reflect more red and less NIR. Every vegetation index is built on this comparison.
- Standard RGB cameras capture useful information – stand counts, drainage patterns, equipment problems, visible stress – but cannot detect early stress or calculate precise vegetation indices.
- Five key wavelength bands matter for crop monitoring: blue, green, red, red edge, and near-infrared. The red edge band is especially sensitive to chlorophyll and nitrogen status.
- Several factors beyond plant health affect what the camera captures: o soil background, sun angle, cloud shadows, canopy architecture, and leaf moisture. Understanding these helps you interpret maps accurately.
- The real power is in combining aerial imagery (which shows you where problems are) with your farming experience (which tells you why). Always ground-truth before making management decisions.

Chapter 4 Notes and References

1. Knipling, E.B. "Physical and Physiological Basis for the Reflectance of Visible and Near-Infrared Radiation from Vegetation." Remote Sensing of Environment, 1(3), 155–159, 1970. The foundational paper explaining why plants reflect NIR.

2. Gates, D.M., Keegan, H.J., Schleter, J.C., and Weidner, V.R. "Spectral Properties of Plants." Applied Optics, 4(1), 11–20, 1965.

3. Tucker, C.J. "Red and Photographic Infrared Linear Combinations for Monitoring Vegetation." Remote Sensing of Environment, 8(2), 127–150, 1979. The paper that introduced NDVI.

4. Jones, H.G. and Vaughan, R.A. Remote Sensing of Vegetation: Principles, Techniques, and Applications. Oxford University Press, 2010.

5. Mulla, D.J. "Twenty-Five Years of Remote Sensing in Precision Agriculture." Journal of Environmental Quality, 42(2), 358–377, 2013.

6. USDA Economic Research Service. "Precision Agriculture in the Digital Age." Economic Information Bulletin No. 248, February 2023. https://www.ers.usda.gov/publications/pub-details?pubid=105893

7. Hatfield, J.L. and Prueger, J.H. "Value of Using Different Vegetative Indices to Quantify Agricultural Crop Characteristics at Different Growth Stages Under Varying Management Practices." Remote Sensing, 2(2), 562–578, 2010.

8. Gitelson, A.A., Viña, A., Arkebauer, T.J., Rundquist, D.C., Keydan, G., and Leavitt, B. "Remote Estimation of Leaf Area Index and Green Leaf Biomass in Maize Canopies." Geophysical Research Letters, 30(5), 2003.

Note: The descriptions of plant reflectance properties in this chapter represent simplified explanations of well-established biophysical principles. For a more technical treatment, see Jones and Vaughan (2010) listed above.

Chapter 5: Vegetation Indices – The Math That Does the Thinking For You

You don't need to understand the formulas. You need to understand what the numbers are telling you about your field. This chapter translates the math into management decisions.

In Chapter 4, we established the fundamental principle: healthy plants absorb red light and reflect near-infrared light, while stressed plants do the opposite. That's the foundation. But when you're staring at a drone map with a rainbow of colors splashed across your field, you need something more specific than a general principle. You need to know what those colors mean, how much you should trust them, and what—if anything—you should do about them.

That's where vegetation indices come in. A vegetation index is simply a formula that takes the light values captured by your camera and converts them into a single number for each point in your field. That number tells you something about the plant at that location—how much chlorophyll it has, how vigorous it is, whether it's stressed. String all those numbers together across your field and you get a map. The map turns abstract light data into a picture you can actually use.

The good news is that you don't need to do any of this math yourself. Your drone's software handles the calculations automatically. What you need is the ability to look at the output and understand what you're seeing. That's what this chapter is for.

We'll cover the five indices you're most likely to encounter, what each one measures, when to use it, and—just as important—when its limitations make it unreliable. By the end of this chapter, you'll know which index to look at for any given situation and what the numbers actually mean for your corn, soybeans, or wheat.

Before We Start: Reading the Color Scale

Every vegetation index map uses a color scale to represent values. Before we discuss individual indices, let's make sure you understand how to read these maps, because the colors can be confusing if you've never seen one before.

Most crop health maps use some version of a red-to-green color scheme. Low values—bare soil, dead plants, severe stress—show up in red or brown. Medium values—moderate vegetation or mild stress—appear in yellow or orange. High values—healthy, vigorous crop—display as green or dark green. Some software uses a more detailed rainbow scale with blues and purples at the extremes.

The key thing to understand is that these colors are relative. The software assigns colors based on the range of values in your specific image. A "yellow" area on one map might represent a value of 0.55, while a "yellow" area on a different map—or even the same field on a different day—might represent 0.45. Always check the legend or scale bar on your map before interpreting colors. The absolute numbers matter more than the colors.

IMPORTANT: COLORS ARE RELATIVE

Don't compare colors between different maps unless you've set them to the same scale. A "green" field in one image and a "yellow" field in another might actually have similar index values—the software just stretched the color scale differently. Always look at the numbers, not just the colors. Most software lets you set a fixed scale for consistent comparisons across dates.

With that foundation, let's look at the indices themselves. We'll start with the one you'll hear about most often.

NDVI: The Workhorse Index

What it stands for: Normalized Difference Vegetation Index

What it measures: Overall plant vigor and green biomass. NDVI is the most widely used vegetation index in agriculture—and in remote sensing in general. It's been around since the 1970s, and there's a reason it's still the default on most mapping software: for a broad-brush assessment of crop health, it works.

The Formula (You Don't Need to Memorize It)

NDVI takes the principle from Chapter 4—healthy plants absorb red light and reflect NIR—and turns it into a simple ratio:

NDVI FORMULA

NDVI = (NIR – Red) / (NIR + Red)

That's it. The software measures how much near-infrared light and how much red light is reflecting from each point in your field, plugs those values into this formula, and gives you a number between -1 and +1.

A healthy plant absorbs most of the red light (Red is low) and reflects lots of NIR (NIR is high). Plug that in: big number minus small number, divided by big number plus small number. You get a value close to 1. A stressed or dead plant reflects similar amounts of red and NIR. Plug that in: the difference is small, so the result is close to 0.

You will never need to do this calculation by hand. The drone's software does it for every pixel in your image – potentially millions of calculations in seconds. What you need is to understand what the output means.

What the Numbers Mean

Value Range	Map Color	What It Means
Below 0.2	Red/Brown	Bare soil, water, rocks, dead vegetation, or pavement. No significant living plant material.
0.2 – 0.4	Orange/Yellow	Sparse vegetation or very stressed plants. Early-season crops before canopy development, or severely stressed areas.
0.4 – 0.6	Yellow/Light Green	Moderate vegetation. Could indicate a developing crop that hasn't reached full vigor, or an established crop under stress.
0.6 – 0.8	Green	Healthy, vigorous crop with good chlorophyll content and active photosynthesis. This is your target zone.
0.8 – 1.0	Dark Green	Very dense, healthy vegetation with full canopy closure. Maximum biomass and vigor.

For Midwest row crops during the growing season, you'll typically see values between 0.3 (early season, sparse canopy) and 0.85 (full canopy, healthy crop). Don't panic if your entire field isn't hitting 0.9—very few fields do outside of a research plot. What matters is consistency within a field. If most of your corn is reading 0.72 and a patch is reading 0.48, that patch needs your attention.

Best Uses for NDVI

Early-season stand assessment. At V4–V6, before the canopy closes, NDVI clearly distinguishes between areas with good plant populations and areas with poor emergence or skips. The contrast between green plants and bare soil makes problem areas obvious.

General health screening. NDVI is your first-pass tool for identifying areas that need closer inspection. It won't tell you why a spot is stressed, but it will tell you where to look.

Season-over-season comparisons. Because NDVI has been the standard for decades, there's a deep baseline of reference data. You can compare your field's NDVI at V8 this year to the same field at V8 last year and see whether conditions are tracking better or worse.

Yield estimation. Research has shown moderate correlations between mid-season NDVI values and final yield, particularly in corn. It's not a crystal ball, but it gives you a rough sense of relative productivity across a field.

The Saturation Problem: NDVI's Big Limitation

Here's where NDVI falls short, and it's important enough to warrant its own section.

NDVI "saturates" at high biomass levels. What does that mean in plain language? Once your crop reaches full canopy closure—when the leaves cover all the soil and the camera sees nothing but green from above—NDVI values plateau. A field with healthy corn at VT might read 0.82. A field with exceptionally healthy, perfectly fertilized corn at VT might read 0.84. The difference is tiny, even though the actual difference in crop health and yield potential could be meaningful.

Think of it like a thermometer that only goes to 100 degrees. It works great for telling you whether it's 60 or 80 outside. But if it's 102 or 108, the thermometer just reads 100 both times. You've lost the ability to distinguish between "hot" and "really hot."

This saturation effect means NDVI is most useful early to mid-season, before full canopy closure. Once the canopy closes – typically around V10–VT in corn – NDVI becomes less helpful for detecting the kind of subtle stress differences that drive management decisions. It can still pick up severe problems (dead plants are always obvious), but it struggles with the gradations that matter most.

This is exactly why NDRE was developed.

When to Use NDVI

Crop	Best Window	Growth Notes
Corn	V4 through VT	Early to mid-season. Most useful before canopy closure; loses sensitivity at full canopy.
Soybeans	V3 through R3	Early to mid-season. Canopy closes earlier than corn; switch to NDRE sooner.
Wheat	Greenup through flag leaf	Spring greenup through heading. Useful for winter kill assessment and early vigor mapping.

NDRE: The Nitrogen Detective

What it stands for: Normalized Difference Red Edge

What it measures: Chlorophyll content in leaves, which is directly related to nitrogen status. If NDVI is your general health screening, NDRE is the specialist you call when you need a more nuanced diagnosis.

Why NDRE Exists

NDRE was developed specifically to address the saturation problem we just discussed with NDVI. It works on the same basic principle – comparing reflected light at two different wavelengths – but swaps out the red band for the red-edge band.

Remember the red edge from Chapter 4? It's that narrow transition zone between visible red and near-infrared where plant reflectance changes rapidly. The red-edge band is exquisitely sensitive to chlorophyll concentration. Even when the canopy is fully closed and NDVI has hit its ceiling, the red-edge band can still detect differences in chlorophyll content from leaf to leaf and area to area.

NDRE FORMULA

NDRE = (NIR – Red Edge) / (NIR + Red Edge)

Same structure as NDVI, just using the red-edge band instead of the red band. The software handles all of this – you just need to know which map to pull up.

Why Chlorophyll Equals Nitrogen

The reason NDRE is so useful for nitrogen management is a direct biological link: nitrogen is a key building block of chlorophyll. When a corn plant has adequate nitrogen, it produces plenty of chlorophyll, and its leaves are dark green with strong red-edge reflectance characteristics. When nitrogen is deficient, the plant can't produce as much chlorophyll. The red-edge signal shifts. NDRE picks up that shift.

This is essentially what happens when corn "fires"—the lower leaves turn yellow because the plant is cannibalizing chlorophyll from older leaves to feed the developing ear. By the time you see that yellowing, the nitrogen shortage has been affecting the plant for weeks. NDRE can flag the shortage earlier, while there's still time to intervene with a side-dress application.

What the Numbers Mean

Value Range	Map Color	What It Means
Below 0.2	Red/Brown	Severe stress, bare ground, or dead vegetation. Immediate ground-truthing is needed.
0.2 – 0.3	Orange	Significant stress, likely nitrogen deficiency. A strong candidate for supplemental nitrogen application.
0.3 – 0.4	Yellow	Below-average chlorophyll. This could indicate developing deficiency — monitor closely and compare to field history.
0.4 – 0.5	Green	Adequate chlorophyll content. The crop is generally healthy with reasonable nitrogen supply.
Above 0.5	Dark Green	Strong chlorophyll content. A healthy crop with good nitrogen uptake. No intervention needed.

Note that NDRE values run lower than NDVI values for the same field. Don’t look at an NDRE map showing 0.45 and compare it mentally to an NDVI map showing 0.75—they’re different scales measuring different things. A healthy full-canopy cornfield might show NDVI of 0.80 and NDRE of 0.50 at the same time. Both are normal.

Best Uses for NDRE

Mid-season nitrogen assessment. This is NDRE's sweet spot. At V8–V12, when you're making side-dress decisions, NDRE can identify areas within a field that are running short on nitrogen while there's still time to act. This is the single most financially impactful use of drone-based monitoring for most Midwest corn operations.

Late-season stress detection. After canopy closure, when NDVI has plateaued, NDRE continues to detect differences. Disease onset, late-season nitrogen burndown, and drought stress all show up in NDRE when NDVI might miss them.

Variable-rate nitrogen prescriptions. If you're set up for variable-rate application, NDRE maps at V8–V10 can be translated into zone-based prescriptions that put more nitrogen where it's needed and less where it's not. We'll cover this process in detail in Chapter 15.

The Catch: You Need a Multispectral Camera

Here's the trade-off. NDRE requires a camera that can capture the red-edge wavelength – roughly 700–740 nanometers. Standard RGB cameras can't do this. You need a multispectral sensor. That's a significant investment (see Chapter 7 for the cost-benefit analysis).

The alternative: hire a service provider to fly NDRE for you two or three times during the season. At a typical cost of $8–$15 per acre for a few hundred acres, it's a fraction of the equipment cost, and you get professionally processed maps without the learning curve. Chapter 17 covers how to find and work with service providers.

When to Use NDRE

Crop	Best Window	Growth Notes
Corn	V8 through R3	Mid to late season. Peak value at side-dress timing (V8–V12); doesn't saturate at full canopy.
Soybeans	R3 through R5	Mid to late season. Useful for detecting late-season stress and disease pressure.

Crop	Best Window	Growth Notes
Wheat	Flag leaf through grain fill	Spring through grain fill. Nitrogen status during critical growth; supports fungicide timing decisions.

GNDVI: A Different Angle on Chlorophyll

What it stands for: Green Normalized Difference Vegetation Index

What it measures: Chlorophyll concentration, similar to NDVI but using the green band instead of the red band. Think of it as NDVI's cousin that's slightly more sensitive to chlorophyll levels.

GNDVI FORMULA

GNDVI = (NIR – Green) / (NIR + Green)

Same structure as NDVI and NDRE. The only difference is which visible band is paired with NIR.

Why GNDVI Matters

Green light penetrates deeper into the leaf canopy than red light does. Because of this, GNDVI is more sensitive to chlorophyll concentration in the middle and lower portions of the canopy—areas where nitrogen stress often shows up first. It's also less prone to saturation than standard NDVI, though not quite as resistant to saturation as NDRE.

GNDVI occupies a useful middle ground. It can be calculated from any multispectral camera that captures green and NIR bands – which is all of them. It offers better chlorophyll sensitivity than NDVI without requiring the red-edge band that NDRE needs. For some cameras and software packages that don't include a dedicated red-edge channel, GNDVI is the best available option for chlorophyll and nitrogen assessment.

Best Uses for GNDVI

Early-season chlorophyll assessment. At V4–V8, before full canopy closure, GNDVI provides a more sensitive read on chlorophyll status than NDVI. If you're monitoring nitrogen uptake early in the season, GNDVI may reveal differences that NDVI misses.

Nitrogen status when NDRE isn't available. If your multispectral camera lacks a red-edge band, or if your software only offers GNDVI as a chlorophyll-related index, it's a solid alternative. Not as precise as NDRE for mid-season nitrogen work, but better than NDVI.

Comparing chlorophyll across different soil backgrounds. Because the green band interacts with vegetation differently than the red band, GNDVI can sometimes provide a cleaner signal in fields with highly variable soil types, especially early in the season.

Most farmers who have access to NDRE will reach for NDRE instead of GNDVI for nitrogen work – and that's the right call. But GNDVI is worth knowing about because you'll see it as an option in most mapping software, and there are situations where it earns its place.

SAVI: The Early-Season Specialist

What it stands for: Soil-Adjusted Vegetation Index

What it measures: Plant vigor with a built-in correction for soil brightness. SAVI addresses a specific problem: when your camera can see a lot of bare soil between the rows, standard NDVI values get contaminated by the soil signal. SAVI mathematically reduces that contamination.

SAVI FORMULA

SAVI = [(NIR – Red) / (NIR + Red + L)] x (1 + L)

The "L" is a soil brightness correction factor, typically set to 0.5 for most agricultural applications. You don't need to adjust it—the software handles it. The important thing is that this correction reduces the influence of soil background on the final number.

Why SAVI Matters for Midwest Farmers

Western Ohio soils range from dark Brookston silty clay loams with 3–4% organic matter to lighter Miami silt loams with less than 2%. In a field that transitions between these soil types—which is common in our region—an early-season NDVI map will show patterns that look like crop health variability but are actually just soil color differences. Dark soil reflects less light overall, which pushes NDVI values in one direction. Light soil reflects more, pushing values the other way.

SAVI accounts for this. When you fly a V4 cornfield with variable soil types, SAVI gives you a more accurate picture of actual plant health by dialing down the soil noise. It's not perfect—no index is—but it's measurably better than NDVI for early-season work on fields with significant soil variability.

Best Uses for SAVI

Stand assessment at emergence. When evaluating emergence uniformity at V2–V4, SAVI provides a cleaner read on actual plant presence and vigor than NDVI, especially on fields with variable soil types.

Replant decisions. If you're trying to determine whether a poor-emergence area warrants replanting, SAVI gives you a more honest assessment of what's actually growing versus what's just soil color variation.

Fields with heavy residue or variable tillage. No-till fields with varying amounts of corn residue on the surface can fool standard NDVI. SAVI handles this better.

Once the canopy closes and the camera can no longer see the soil, SAVI's advantage disappears. At that point, it produces values very similar to NDVI. Think of SAVI as a specialized tool for the first few weeks of the growing season. After canopy closure, switch to NDVI or NDRE for the remainder of the season.

Chlorophyll Index (CI): Straight to the Source

What it measures: Leaf chlorophyll concentration, as directly as any remote measurement can. While NDVI and NDRE infer chlorophyll status indirectly through the overall reflectance signature, the Chlorophyll Index (CI) is designed to target chlorophyll content specifically.

There are actually several variants of Chlorophyll Index used in agricultural remote sensing. The two most common are:

CI FORMULAS CI Red Edge (CI-RE) = (NIR / Red Edge) – 1

CI Green (CI-G) = (NIR / Green) – 1

Both produce values that correlate strongly with actual chlorophyll content measured in the lab. The red-edge version (CI-RE) tends to be more sensitive at high chlorophyll levels, making it particularly useful for well-fertilized crops.

Why Chlorophyll Index Matters

Chlorophyll is the molecule that drives photosynthesis. More chlorophyll means more photosynthetic capacity means more potential yield. Chlorophyll production requires nitrogen. So chlorophyll content is a direct, biological indicator of whether your crop has the nitrogen it needs to reach its potential.

The Chlorophyll Index approaches this measurement more directly than NDVI or even NDRE. Research from the University of Nebraska and other institutions has shown that CI-RE correlates more closely with actual measured chlorophyll content than either NDVI or NDRE across a wide range of conditions. In practical terms, this means CI can sometimes detect nitrogen deficiency when other indices are ambiguous.

Best Uses for Chlorophyll Index

Nitrogen management decisions. If your mapping software offers a Chlorophyll Index option, it's worth comparing to NDRE for mid-season nitrogen assessment. In some cases – particularly with well-fertilized corn where you're trying to distinguish between "adequate" and "optimal" nitrogen – CI may reveal differences that NDRE doesn't.

Identifying deficiency before visual symptoms. CI picks up declining chlorophyll production at the leaf level. Because chlorophyll loss precedes visible yellowing, CI can provide earlier warning of developing nitrogen stress than visual scouting.

Research and precision applications. If you're working with an agronomist or a university extension specialist on nitrogen rate trials, CI provides a more quantitative measure of crop nitrogen status that can be correlated with tissue samples and soil tests.

For most farmers in their first or second year of drone monitoring, the Chlorophyll Index is a "nice to have" rather than a "must have." NDRE covers most of the same ground with similar accuracy. But as you get more comfortable with your maps and want more precise nitrogen management, CI is a tool worth exploring.

Putting It Together: Which Index for Which Situation

Now that you know the five main indices, the practical question is: which one should I look at, and when? Here's the decision framework, condensed into a reference you can keep in the truck.

Situation	Best Index	Camera	Why
Early-Season Stand Assessment (V2–V6)	SAVI or NDVI	Multispectral (RGB can help visually)	Accounts for soil background; distinguishes plants from bare ground.

Situation	Best Index	Camera	Why
Pre-Canopy General Health Check (V4–V8)	NDVI	Multispectral	Strong contrast between healthy and stressed vegetation before canopy closure.
Side-Dress Nitrogen Decision (V8–V12)	NDRE	Multispectral with red-edge band	Best chlorophyll sensitivity at full canopy; directly related to nitrogen status.
Full-Canopy Health Monitoring (VT–R3)	NDRE or CI	Multispectral with red-edge band	NDVI saturates at full canopy; NDRE and CI continue to detect differences.
Late-Season Stress Detection (R3–R6)	NDRE	Multispectral with red-edge band	Detects disease, drought, and late nitrogen burndown in closed canopy.
Scouting and Visual Problem-Finding	No index needed	RGB (any camera is fine)	Use your eyes on the orthomosaic to spot patterns.
Drainage and Tile Line Mapping	No index needed	RGB (any camera is fine)	Water patterns visible in standard color imagery; no index needed.
Comparing Fields with Different Soil Types	SAVI	Multispectral	Soil brightness correction reduces false variability between fields.

THE PRACTICAL SUMMARY

If you only remember two indices: use NDVI early in the season (before canopy closure) and NDRE from mid-season on. That combination covers 90% of the management decisions most Midwest row crop farmers need to make.

If you're flying with an RGB camera only, skip the indices entirely. Your standard color imagery is your tool – use it for scouting, stand counts, drainage mapping, and identifying visible problems. That's still tremendously valuable.

Reading Your Maps: What the Colors Actually Mean

Let's walk through a practical example of map interpretation, because looking at index values in a table is different from looking at a map of your field.

Imagine you've flown a 160-acre cornfield at V10 with a multispectral camera. Your software generates an NDRE map. The legend shows values from 0.15 to 0.55. Most of the field is green (0.40–0.50). But there's a band of yellow running diagonally across the southwest quarter (0.28–0.35) and a few small red spots scattered along the east edge (0.18–0.22).

Here's how to think about it:

The green areas (0.40–0.50) are your healthy corn. Adequate chlorophyll, good nitrogen uptake, on track for target yield. No action needed in these zones.

The yellow band (0.28–0.35) is telling you chlorophyll levels are below where they should be at V10. This is strong evidence of nitrogen deficiency—especially since it's a band rather than random scattered spots. A band pattern often traces back to an applicator issue (spreader overlap, variable-rate prescription boundary mismatch) or a soil type transition that affects nitrogen availability. This area is a candidate for supplemental nitrogen if you're still within the side-dress window.

The red spots (0.18–0.22) represent severely stressed or possibly dead plants. At these values, something beyond simple nitrogen deficiency is going on. Ground-truth immediately. Could be standing water, herbicide carryover, severe compaction, or seedling disease. These spots may be small enough that the economics of treatment don't justify action, but they're worth understanding for future management.

Notice what we just did: we looked at the map, identified the patterns, formed hypotheses about causes, and made preliminary management decisions—all in about two minutes. But we haven't sprayed anything yet. The next step is ground-truthing: drive out to the yellow band, walk it, look at the plants, maybe pull a tissue sample. Confirm the diagnosis before spending money on treatment. Chapter 10 covers the ground-truthing process in detail.

Common Mistakes in Index Interpretation

As you start working with vegetation indices, watch out for these common errors. Every farmer who's used this technology has made at least one of them.

Comparing Maps on Different Scales

Your software generates an NDVI map for Field 3 on June 10 and another for Field 3 on July 5. The June map looks "more green" than the July map. Does that mean the crop got worse? Not necessarily. Check the legend. The June map might use a scale of 0.30–0.65, while the July map uses 0.55–0.85. The crop actually improved, but the auto-scaled colors make it look like it didn't. Always lock your color scale to a fixed range when comparing dates.

Confusing NDVI and NDRE Scales

NDVI and NDRE produce different value ranges for the same crop. An NDVI of 0.75 and an NDRE of 0.45 might both represent "healthy corn." Don't mix them up in your head or in conversation. Know which index you're looking at before interpreting the numbers.

Chasing Cloud Shadows

We mentioned this in Chapter 4, but it bears repeating: cloud shadows create dark patches on your map that look exactly like crop stress. If your flight was on a partly cloudy day, be suspicious of any stress pattern that doesn't correspond to a known field feature. Check the timestamp on the individual images if your software allows—you may be able to see the shadow moving across frames.

Acting on a Single Flight

One map gives you a snapshot. Two maps a week apart give you a trend. If a stress area shows up on one flight, note it. If it shows up on two flights, take it seriously. Plants can have a bad day—maybe they were temporarily water-stressed after a hot afternoon when you flew, but recovered overnight. Patterns that persist across multiple flights are the ones worth acting on.

Expecting Pixel-Level Accuracy

Vegetation indices give you field-level and zone-level information, not plant-level precision. A single pixel in your map might represent a 3x3 foot area containing multiple plants. Don't obsess over individual pixel values. Look for patterns—zones, bands, gradients—that span multiple pixels and correspond to real-world features. If a "problem" is only one or two pixels, it's probably noise.

Quick Reference: Vegetation Index Summary

This summary is designed for quick reference. A laminated version is included in Appendix A.

Index	Full Name	Formula	Best For	Limitation
NDVI	Normalized Difference Vegetation Index	(NIR – Red) / (NIR + Red)	Early-season health, stand assessment, general screening	Saturates at full canopy; can't distinguish "healthy" from "very healthy"

Index	Full Name	Formula	Best For	Limitation
NDRE	Normalized Difference Red Edge	(NIR – Red Edge) / (NIR + Red Edge)	Mid-season nitrogen, full-canopy health, late-season stress	Requires red-edge band (multispectral camera)
GNDVI	Green Normalized Difference Vegetation Index	(NIR – Green) / (NIR + Green)	Chlorophyll assessment, nitrogen status when NDRE unavailable	Less precise than NDRE for nitrogen; less common in software
SAVI	Soil-Adjusted Vegetation Index	[(NIR – Red) / (NIR + Red + L)] x (1+L)	Early-season on variable soils, stand counts	No advantage after canopy closure; adds complexity
CI	Chlorophyll Index	(NIR / Red Edge) – 1 or (NIR / Green) – 1	Precise chlorophyll and nitrogen measurement; research applications	Less intuitive scale; less common in basic software

Chapter Summary

- Vegetation indices are formulas that convert reflected light values into single numbers representing crop health at each point in your field. The software does the math – you interpret the results.
- NDVI is the most common index and works well for early-season health screening, stand assessment, and general crop monitoring. Its main limitation is saturation: once the canopy closes, it can't distinguish between "healthy" and "very healthy."
- NDRE is the go-to index for nitrogen management and mid-to-late-season monitoring. It detects chlorophyll differences that

NDVI misses after canopy closure. It requires a multispectral camera with a red-edge band.

- GNDVI offers a middle ground between NDVI and NDRE, with better chlorophyll sensitivity than NDVI. Useful when a red-edge band isn't available.
- SAVI corrects for soil background and is most valuable early in the season when bare soil is visible between rows. After canopy closure, it offers no advantage over NDVI.
- Chlorophyll Index (CI) directly targets chlorophyll concentration and is the most precise tool for nitrogen assessment, though less commonly available in basic software packages.
- The practical two-index approach: use NDVI before canopy closure and NDRE after. This covers 90% of management decisions for Midwest row crops.
- Always check the color scale legend before interpreting maps. Colors are relative – the same color can mean different values on different maps.
- Common interpretation mistakes include comparing maps on different scales, confusing NDVI and NDRE values, chasing cloud shadows, acting on single flights, and expecting pixel-level accuracy.
- RGB cameras can't calculate true vegetation indices. If you're flying RGB only, skip the indices and use the visual imagery for scouting and pattern identification – that's still highly valuable.

Chapter 5 Notes and References

1. Tucker, C.J. "Red and Photographic Infrared Linear Combinations for Monitoring Vegetation." Remote Sensing of Environment, 8(2), 127–150, 1979. The original NDVI paper.

2. Gitelson, A.A. and Merzlyak, M.N. "Remote Estimation of Chlorophyll Content in Higher Plant Leaves." International Journal of Remote Sensing, 18(12), 2691–2697, 1997. Foundational work on red-edge indices.

3. Huete, A.R. "A Soil-Adjusted Vegetation Index (SAVI)." Remote Sensing of Environment, 25(3), 295–309, 1988. The original SAVI paper.

4. Gitelson, A.A., Viña, A., Ciganda, V., Rundquist, D.C., and Arkebauer, T.J. "Remote Estimation of Canopy Chlorophyll Content in Crops." Geophysical Research Letters, 32(8), 2005.

5. Hatfield, J.L. and Prueger, J.H. "Value of Using Different Vegetative Indices to Quantify Agricultural Crop Characteristics at Different Growth Stages Under Varying Management Practices." Remote Sensing, 2(2), 562–578, 2010.

6. Datt, B. "A New Reflectance Index for Remote Sensing of Chlorophyll Content in Higher Plants." Journal of Plant Physiology, 154(1), 30–36, 1999.

7. Barnes, E.M., Clarke, T.R., Richards, S.E., et al. "Coincident Detection of Crop Water Stress, Nitrogen Status and Canopy Density Using Ground-Based Multispectral Data." Proceedings of the Fifth International Conference on Precision Agriculture, 2000.

8. Mulla, D.J. "Twenty-Five Years of Remote Sensing in Precision Agriculture." Journal of Environmental Quality, 42(2), 358–377, 2013.

9. Thompson, L. and Puntel, L. "Optimize your Nitrogen Investment with Precision Agriculture Technologies." University of Nebraska–Lincoln CropWatch, November 2021. https://cropwatch.unl.edu/2021/optimize-your-nitrogen-investment-precision-agriculture-technologies/

10. USDA Economic Research Service. "Precision Agriculture in the Digital Age." Economic Information Bulletin No. 248, February 2023. https://www.ers.usda.gov/publications/pub-details?pubid=105893

Note: Vegetation index values cited in this chapter represent typical ranges for Midwest row crops under normal growing conditions. Actual values will vary based on crop species, growth stage, soil conditions, sensor calibration, and atmospheric conditions. Use these ranges as guidelines, not absolute thresholds, and develop field-specific benchmarks over multiple seasons.

Chapter 6: Cameras – RGB and Multispectral

Before you spend $5,000 on a multispectral sensor, let's talk about everything you can accomplish with the camera that came with your drone. The answer might surprise you.

If there's one question that comes up more than any other when farmers start looking at drone technology, it's this: "Do I need a multispectral camera, or can I use a regular one?"

The drone industry would love you to believe you need the expensive option. Multispectral sensors, NDVI maps, vegetation indices – the marketing makes it sound like anything less than a $5,000 camera system is a waste of time. And if you've read Chapter 4 and Chapter 5, you might be feeling the pull toward multispectral yourself. After all, near-infrared detection and red-edge sensitivity sound like exactly what you need.

But here's the reality that often gets lost in the sales pitch: a standard RGB camera—the kind that comes built into every consumer drone—can accomplish a remarkable amount of useful agricultural monitoring. Not everything. But far more than most people realize. And for many farmers, particularly those in their first year or two of drone use, RGB is not just "good enough." It's the right starting point.

This chapter makes the case for both sides. We'll catalog everything an RGB camera can do, be honest about what it can't, tackle the "enhanced RGB" question that confuses a lot of buyers, and give you a clear framework for deciding which camp you fall into. By the end, you'll know exactly what you can accomplish with the equipment you already own—and whether it's worth investing more.

What RGB Cameras Can Do: The 80% Solution

RGB stands for Red, Green, Blue—the three color channels that make up every standard photograph. It's the same technology in your phone camera, your truck's dashcam, and the camera built into your DJI Mini. Nothing exotic. Nothing expensive. And in the hands of a farmer who knows what to look for, remarkably powerful.

Let's walk through the specific applications where a standard color camera does the job – often just as well as anything multispectral.

Stand Counts and Emergence Assessment

This is arguably the single highest-value application of any drone camera, and RGB handles it beautifully. At V2–V4, when your corn or soybeans are just a few inches tall, each plant is a distinct green dot against the brown or residue-covered soil. From 200 feet up, a high-resolution RGB camera captures every one of them.

What you can see: missing plants (planter skips), double-planted areas, rows that didn't emerge evenly, sections where crusting or ponding killed seedlings. Some mapping software can even count plants automatically from RGB imagery, giving you an actual population per acre without walking the rows.

Why this matters financially: the replant decision. If a hailstorm or a late frost hammered part of your field, you need to know whether the remaining stand justifies keeping it or whether replanting makes economic sense. Standing in the end rows and guessing is how most farmers handle this. Flying the field at 250 feet and counting what's actually there turns a gut decision into a data decision. At 80 acres of replant cost versus 80 acres of reduced yield, the stakes are high enough to warrant accurate information.

Drainage and Tile Line Mapping

If you do nothing else with your drone, fly your fields after the next significant rain event. What you'll see from 250 feet will change how you think about your drainage – guaranteed.

RGB cameras capture drainage patterns with stunning clarity. Functioning tile lines show up as dark-green strips of healthy, well-drained crop running through lighter or yellower areas where water is slower to move. Broken tiles reveal themselves as abrupt transitions—a healthy strip that suddenly ends, with a wet or stressed zone spreading downstream of the break. Surface drainage patterns, grassed waterway performance, low spots that hold water—all visible in standard color.

One Ohio farmer we spoke with estimated that a single drone flight after a June rain saved him $3,000 in tile repair costs. He was planning to hire a contractor to locate a suspected tile break using a camera system. Instead, he flew the field himself for free, identified the break within ten minutes, and guided the backhoe operator directly to the spot. The repair took two hours instead of the full day it would have taken to locate the problem from the ground.

Weed Pressure and Herbicide Performance

Before canopy closure, weeds between rows are visible from above in standard color imagery. Patches of herbicide-resistant waterhemp, marestail, or giant ragweed stand out against the regular pattern of crop rows – they're in the wrong place, and from altitude, the irregularity is obvious.

More importantly, drone imagery lets you map weed patches and create spot-spray plans. Instead of blanket-treating a 160-acre field because you saw waterhemp in one corner, you can identify the 15 acres that actually need treatment and skip the other 145. At $10–$20 per acre for a post-emergence herbicide pass, that's $1,450–$2,900 in savings on a single application. Even if the drone flight only prevents one unnecessary whole-field spray per season, it's paid for the equipment.

RGB is also excellent for evaluating herbicide performance after application. Fly the field a week after spraying, and you can see exactly where the herbicide worked and where it didn't – information that helps you adjust rates, switch products, or plan follow-up applications.

Equipment Problem Detection

Some of the most financially impactful discoveries farmers make with drones are equipment problems they didn't know they had. RGB imagery reveals these with perfect clarity:

Planter issues. Skips show up as gaps in the row pattern. Doubles appear as thicker rows. Inconsistent planting depth creates alternating bands of better and worse emergence – obvious from 250 feet, nearly invisible from the ground. One misadjusted row unit on a 16-row planter can cost you yield across every field you plant before you catch it.

Fertilizer spreader problems. Overlap and skip patterns from dry fertilizer spreaders create streaks perpendicular to the direction of travel. If your spreader isn't calibrated right or the wind shifted during application, you'll see parallel bands of better and worse crop performance all season long. From the road, it looks like normal variability. From the air, the source is unmistakable.

Sprayer issues. Plugged nozzles create narrow strips of untreated crop. Boom height problems cause uneven coverage. GPS guidance errors create overlap or skip bands. All of these show up in RGB imagery as regular, repeating patterns aligned with the direction of application.

Harvest equipment. If you fly early the following season, you can often see patterns from the previous year's harvest – combine header overlap areas, grain cart traffic compaction, chaff distribution patterns from the combine. These help you understand yield variability that might otherwise seem random.

Damage Assessment and Documentation

When bad things happen to good crops, documentation matters. Hail damage, wind damage, flooding, wildlife destruction – insurance adjusters and FSA offices want evidence, and "it looked pretty bad" doesn't carry the same weight as a timestamped aerial photograph showing exactly how many acres were affected.

RGB drone imagery provides clear, defensible documentation. You can show the extent of a hail swath, the boundary of a flooded area, the acreage impacted by deer damage. The imagery includes GPS coordinates and timestamps automatically. Several crop insurance companies have started accepting drone imagery as supporting evidence for claims, and the trend is moving toward broader acceptance.

Beyond insurance, damage documentation helps with your own record-keeping. Comparing this year's flood damage boundary to last year's helps you make better decisions about drainage investment. Tracking deer damage year over year supports the case for management strategies. The visual record becomes part of your field history.

Visible Stress Identification

Let's be clear about what "visible stress" means in this context. If you can see a problem standing in the field—yellowing, browning, wilting, stunted growth, leaf curling—your RGB camera can see it too. And from 250 feet, it can see it across the entire field simultaneously, revealing patterns and boundaries that ground-level observation misses.

Nitrogen deficiency that's progressed to visible yellowing. Iron chlorosis on calcareous soils. Sulfur deficiency showing as pale, stunted areas. Herbicide injury creating curled or discolored patches. Flooding damage. Frost pockets. All of these show up in standard color imagery once they've reached the point of visible symptoms.

The limitation, as we discussed in Chapter 4, is timing. RGB captures stress once it's visible—which means the damage is already done. Multispectral can catch it earlier. But "earlier" doesn't always mean "actionable." If the growing season has progressed past the point where intervention is practical, catching the stress two weeks earlier wouldn't have changed your decision. Context matters.

General Field Overview and Pattern Recognition

Never underestimate the value of simply seeing your entire field from above. Patterns that are invisible from the ground—or that would take hours of walking to piece together—become immediately obvious from altitude.

Old fence rows buried under decades of tillage. Soil type transitions that track with subtle crop performance differences. Compaction zones from repeated equipment traffic. The edge effects where your neighbor's spray drift reaches your field. The spot where the previous owner buried construction debris that's been stunting your yields for years.

A single RGB flight at mid-season creates a complete visual record of your field's performance that you can overlay with yield maps, soil maps, and as-applied maps. This kind of cross-referencing is how farmers build the deep understanding of field variability that drives profitable management decisions year after year.

What RGB Cameras Can't Do: The Honest Limitations

Now for the other side. Understanding what RGB can't do is just as important as appreciating what it can, because it prevents you from expecting more than the technology can deliver—and helps you know when it's time to upgrade.

Detect Stress Before Visible Symptoms

This is the fundamental limitation, and there's no way around it. A standard camera sees exactly what your eyes see, just from a better vantage point. The near-infrared reflectance changes that signal early stress—the one-to-three-week early warning window we discussed in Chapter 4—are completely invisible to an RGB sensor.

For many applications, this is acceptable. Stand counts don't need early stress detection. Drainage mapping doesn't need it. Equipment problems are visible in RGB. But for the most financially impactful use case – catching nitrogen deficiency at V8 before it becomes visible at V12 – RGB falls short.

Calculate Accurate Vegetation Indices

Without a near-infrared channel, true vegetation indices like NDVI and NDRE simply cannot be calculated. The formulas require NIR data. No amount of software processing can extract information that the sensor never captured.

Some software will generate what they call a "RGB-NDVI" or "visible-band vegetation index" using only the red and green channels from a standard camera. These can reveal relative patterns—the software is essentially mapping color differences—but the values are not true NDVI, they're not calibrated, and they shouldn't be used for precise prescription-making. We'll address this in more detail shortly.

Provide Precise Nitrogen Assessment

Nitrogen management is where multispectral earns its keep. The red-edge band – essential for NDRE – is specifically tuned to chlorophyll content, which directly reflects nitrogen status. RGB cameras have no access to this wavelength. You can see severe nitrogen deficiency once it causes yellowing, but the subtle, within-field variability that drives variable-rate prescriptions is beyond RGB's capability.

Detect Early Disease

Disease detection from drones is limited even with multispectral cameras, as we'll discuss in Chapter 15. But to the extent that drone-based disease detection works, it relies on changes in NIR reflectance and red-edge response that precedes visible lesion development. RGB can only show you disease after it's visible – and by then, you've likely already noticed it during ground scouting.

Distinguish Between Types of Stress

This is a subtler limitation. When an RGB camera shows a yellow patch, it's telling you "something is wrong here." But it can't easily distinguish between nitrogen deficiency (yellowing), sulfur deficiency (pale green), iron chlorosis (interveinal yellowing), and several other stresses that all produce similar visible symptoms. Multispectral cameras, by capturing additional wavelength bands, provide more information that helps narrow the diagnosis—though they can't make the final call either. Ground-truthing is always required.

Side by Side: RGB vs. Multispectral Capabilities

Here's a comprehensive comparison of what each camera type can accomplish for common agricultural applications:

Application	RGB Camera	Multispectral
Stand counts and emergence assessment		
Drainage and tile line mapping		
Weed patch identification		
Equipment problem detection		
Damage assessment and documentation		
Visible stress identification		
General field overview and patterns		
Early stress detection (before visible)	[ ]	
True NDVI calculation	[ ]	
NDRE / nitrogen assessment	[ ]	
Calibrated vegetation indices	[ ]	

Application	RGB Camera	Multispectral
Variable-rate prescription creation	[ ]	
Chlorophyll content mapping	[ ]	
Enhanced RGB / approximate indices	N/A	~
= Yes	[ ] = No	~ = Limited/approximate

Count the checkmarks. RGB handles seven of the thirteen applications listed. The six it can't handle are all related to the same underlying limitation, which is no near-infrared capability. If those six applications matter to your operation—and for many larger farms, they do—multispectral is the path. But if you're primarily scouting, assessing stands, mapping drainage, and documenting conditions, RGB covers the ground.

The "Enhanced RGB" Question: Can Software Bridge the Gap?

If you've spent any time shopping for drone mapping software, you've probably encountered claims like "Generate NDVI maps from any camera!" or "Precision agriculture analysis from RGB imagery." This is one of the most confusing areas in the consumer drone market, so let's untangle it.

What "Enhanced RGB" Actually Does

Several software platforms can take standard RGB photographs and generate maps that look like NDVI maps – complete with the red-to-green color scheme and numerical values. What they're actually doing is calculating a visible-band vegetation index, most commonly something called VARI (Visible Atmospherically Resistant Index) or ExG (Excess Green), and presenting it in a format that mimics true NDVI.

These visible-band indices use the mathematical relationships between the red, green, and blue channels to estimate plant vigor. The logic is sound at a basic level: healthy plants reflect more green and less red in the visible spectrum, so a ratio of green-to-red from a standard camera does carry some information about plant health.

What It Gets Right

Enhanced RGB indices can reveal relative variability within a field. If one area of your field is significantly less vigorous than another, a visible-band index will often flag it. The patterns it shows – stressed zones, healthy zones, transitions – are frequently consistent with what a multispectral camera would reveal.

For basic scouting prioritization—identifying which areas of the field need ground-truthing—enhanced RGB can be useful. It's essentially automating what your eyes would do if they could look at the whole field simultaneously: spotting the areas that look different.

What It Gets Wrong

The problems start when you try to use enhanced RGB data for quantitative decisions. The values aren't calibrated to any physical measurement. An enhanced RGB "NDVI" of 0.6 doesn't mean the same thing as a true NDVI of 0.6 from a multispectral camera. You can't compare values between dates reliably, because changes in lighting conditions shift the entire scale. You can't create accurate variable-rate prescriptions from the data. And you definitely can't detect stress before it's visible—the visible bands simply don't carry that information.

Enhanced RGB indices are also highly sensitive to lighting conditions. The same field on the same day can produce different results depending on whether the sky was overcast or sunny, what time you flew, and whether a cloud passed over during the flight. True multispectral systems account for this through calibration panels and light sensors; RGB has no such correction.

THE BOTTOM LINE ON ENHANCED RGB

Think of it this way:

Enhanced RGB is like a home blood pressure cuff. It gives you a general sense of whether something might be off, and it's useful for spotting trends. But you wouldn't make a medical decision based solely on its reading – you'd confirm with clinical equipment.

Use enhanced RGB maps to identify areas that warrant ground-truthing. Don't use them to write variable-rate prescriptions or make precise input decisions. If the pattern matters enough to act on, confirm it with boots on the ground.

What I Spotted With Just an RGB Camera: Real Examples

Theory is one thing. Seeing real results is another. Here are examples of valuable discoveries made with nothing more than a standard RGB camera on a consumer drone—all from farms in the Western Ohio and Eastern Indiana region.

The $8,000 Planter Fix

A Preble County farmer flew his corn at V4 after noticing a few gaps during a drive-by. The RGB orthomosaic showed a systematic problem: every eighth row had noticeably poorer emergence. The pattern repeated across all 200 acres he'd planted with that planter. Ground inspection revealed a worn seed tube on row unit eight that was occasionally dropping two seeds or no seeds instead of singulating properly. Cost to repair: $85 for a new seed tube. Estimated yield saved across the remaining fields yet to be planted: over $8,000.

The Hidden Waterway

A Montgomery County farm manager had always wondered why a particular area of a 120-acre field consistently underperformed. Yield maps showed a subtle dip, but nothing pointed to a cause. A post-rain RGB flight in early June revealed a surface drainage pattern he'd never seen from the ground: water was channeling along a shallow swale that had been graded over decades ago but still functioned as a waterway during heavy rainfall. The crop along this path was stunted from repeated inundation. The solution—a grass waterway and grading adjustment—addressed a problem that had been costing him an estimated 5–8 bushels per acre on about 10 acres every year.

The Neighbor's Drift

A Darke County soybean grower noticed some odd curling on plants near his western field boundary. A quick RGB flight revealed the extent: a strip approximately 80 feet wide running the entire length of the shared boundary—roughly 3 acres of beans showing herbicide drift symptoms from his neighbor's corn herbicide application. The timestamped, geotagged drone imagery provided clear documentation for the conversation with his neighbor and, ultimately, for the insurance claim.

The Compaction Map

After harvest, a Wayne County farmer flew a freshly planted wheat field in late October. The RGB imagery showed a distinctive pattern: lighter-colored strips where the wheat was emerging more slowly, running parallel to the previous corn harvest. The strips matched exactly where the combine and grain cart had run. The compaction from fall harvest was delaying wheat emergence in the traffic rows. The farmer marked the worst areas for spring tillage and began planning a controlled traffic pattern for the following year's harvest.

Stand Count After the Storm

A late-May hailstorm damaged portions of three fields on a Miami County operation. The farmer flew all three fields the next morning using his DJI Mini 4 Pro. The RGB imagery clearly showed the hail swath – a band roughly 400 yards wide cutting diagonally across the fields. Within the swath, plant density was visibly reduced. Outside it, emergence looked normal. The imagery allowed him to work with his crop insurance adjuster to delineate the damaged area precisely: 47 acres total across the three fields, rather than the roughly 80 acres he'd initially estimated from driving the roads. The accuracy saved him from replanting 33 acres that didn't need it – a decision worth approximately $6,000 in avoided replant costs.

Decision Guide: When Is RGB Enough?

Here's the practical framework. Answer these questions to determine whether RGB meets your needs or whether multispectral would provide meaningful additional value.

RGB IS LIKELY ENOUGH IF...

- *You're in your first or second year of drone use and still learning the basics*
- *Your primary goals are scouting, stand assessment, drainage mapping, and damage documentation*
- *You manage fewer than 500 acres of row crops*
- *You don't currently use variable-rate application equipment*
- *Your fields are relatively uniform in soil type and productivity*
- *You want to prove the value of aerial monitoring before committing to a larger investment*

CONSIDER MULTISPECTRAL IF...

- *You manage 800+ acres and nitrogen is a major input cost*
- *You already own or have access to variable-rate application equipment*
- *Your fields have significant variability in soil type, drainage, or productivity*
- *You're already making side-dress nitrogen decisions and want better data to guide them*
- *You're comfortable with your basic drone skills and ready for more advanced applications*
- *You're offering (or planning to offer) drone services to other farmers*

Notice something important: neither list mentions the technology itself. The decision isn't really about cameras and sensors. It's about what you're trying to accomplish, the scale of your operation, and whether you're in a position to act on the additional information multispectral provides. The most expensive multispectral camera in the world is worthless if you don't have variable-rate equipment to implement the prescriptions it generates.

Cost-Benefit Comparison: RGB Only vs. Multispectral Investment

Let's put real numbers on this decision. The following comparison assumes a 600-acre corn and soybean operation in Western Ohio.

Option A: RGB Only

Item	Cost	Notes
DJI Mini 4/5 Pro Fly More Combo	$1,050–$1,280	Three batteries, charging hub
Accessories	$75–$140	microSD cards, landing pad, ND filters
Part 107 Certification	$175	One-time, valid two years
Software (free tier)	$0	DJI Fly app, OpenDroneMap
Annual Operating Cost	$200–$400	Batteries, insurance endorsement

Total Year 1 = ~ $1,500–$2,000

Annual Cost Thereafter = ~ $200–$400

Option B: Multispectral System

Item	Cost	Notes
DJI Mavic 3 Multispectral	$4,959+	Integrated multispectral + RGB

Item	Cost	Notes
Accessories and Extras	\$300–\$500	Batteries, calibration panel
Part 107 Certification	\$175	One-time, valid two years
Processing Software	\$1,650–\$3,500+/yr	Pix4Dfields to DroneDeploy
Annual Operating Cost	\$500–\$800	Batteries, insurance, calibration

Total Year 1 = ~ \$6,300–\$8,900

Annual Cost Thereafter = ~ \$2,150–\$3,200 (depending on software subscription)

Option C: The Hybrid Approach

Item	Cost	Notes
GB Drone (owned)	\$1,050–\$1,280	For scouting, stands, drainage
Hire Multispectral Service (three flights)	\$2,400–\$5,400	600 acres x \$8–\$15/acre x three flights
Part 107 Certification	\$175	One-time

Total Year 1 = ~ \$3,625–\$6,855

Annual Cost Thereafter = ~ \$2,400–\$5,400 (service flights only)

The hybrid approach—owning an RGB drone for frequent scouting flights while hiring a multispectral service for two or three key flights per season—is increasingly popular, and for good reason. You get the flexibility of daily access for scouting and stand counts (RGB), plus the precision of calibrated NDRE maps at the critical nitrogen decision points (multispectral), without the full capital investment or learning curve of owning and operating a multispectral system.

Chapter 7 digs deeper into the multispectral investment decision, including break-even calculations by acreage. Chapter 17 covers how to find and work with service providers.

The 80/20 Rule of Drone Monitoring

Here's the principle we keep coming back to throughout this book: 80% of the practical value from drone-based crop monitoring comes from applications that a standard RGB camera handles just fine. Stand counts. Drainage mapping. Equipment problem detection. Weed scouting. Damage documentation. Field overview and pattern recognition.

The remaining 20% of value – early stress detection, precise nitrogen assessment, calibrated vegetation indices, variable-rate prescriptions – requires multispectral capability. For a large, sophisticated operation, that 20% can translate into thousands of dollars in additional savings. For a smaller operation or a farmer just getting started, the 80% is more than enough to justify the investment and deliver a positive return.

KEY PRINCIPLE

Start with RGB. Master the basics. Prove the value on your own operation. Then—and only then—decide if the additional 20% of capability justifies the additional cost. You'll make a better investment decision because you'll know exactly what you need, based on experience rather than marketing.

Chapter Summary

- RGB cameras handle approximately 80% of practical agricultural drone applications: stand counts, drainage mapping, weed scouting, equipment problem detection, damage assessment, visible stress identification, and general field overview.
- The 20% that RGB can't do – early stress detection, true NDVI/NDRE, precise nitrogen assessment, and calibrated prescription creation – all require near-infrared capability that only multispectral cameras provide.
- "Enhanced RGB" software can generate approximate vegetation maps from standard cameras. These are useful for identifying patterns and

prioritizing scouting but should not be used for quantitative decisions or prescription creation.

- Real-world examples demonstrate that RGB-only drone flights routinely uncover equipment problems, drainage issues, pest pressure patterns, and damage boundaries worth thousands of dollars in savings or improved management.
- RGB is likely enough if you're new to drones, manage fewer than 500 acres, don't use variable-rate equipment, or want to prove the concept before investing further.
- Consider multispectral if you manage 800+ acres, have variable-rate equipment, need precise nitrogen management, or plan to offer services to others.
- The hybrid approach—owning an RGB drone for frequent flights while hiring multispectral services for key growth stages—offers a practical middle path that balances cost, capability, and learning curve.
- The best investment decision comes from experience: start with RGB, learn what you need, then upgrade when you've identified a specific gap that justifies the cost.

Chapter 6 Notes and References

1. DJI. "DJI Mini 4 Pro Specifications." https://store.dji.com/product/dji-mini-4-pro. Accessed February 2026.

2. DJI. "DJI Mavic 3 Multispectral." https://enterprise.dji.com/mavic-3-m. Accessed February 2026.

3. Gitelson, A.A., Kaufman, Y.J., Stark, R., and Rundquist, D. "Novel Algorithms for Remote Estimation of Vegetation Fraction." Remote Sensing of Environment, 80(1), 76–87, 2002. Foundational work on visible-band vegetation indices.

4. Hunt, E.R., Doraiswamy, P.C., McMurtrey, J.E., Daughtry, C.S.T., Perry, E.M., and Akhmedov, B. "A Visible Band Index for Remote Sensing Leaf Chlorophyll Content at the Canopy Scale." International Journal of Applied Earth Observation and Geoinformation, 21, 103–112, 2013.

5. Woebbecke, D.M., Meyer, G.E., Von Bargen, K., and Mortensen, D.A. "Color Indices for Weed Identification Under Various Soil, Residue, and Lighting Conditions." Transactions of the ASAE, 38(1), 259–269, 1995.

6. USDA Risk Management Agency. "Use of Unmanned Aerial Systems (UAS) in Crop Insurance." Policy guidance, updated 2024.

7. Mulla, D.J. "Twenty-Five Years of Remote Sensing in Precision Agriculture." Journal of Environmental Quality, 42(2), 358–377, 2013.

8. Ohio State University Extension. "Precision Agriculture and Drone Technology for Ohio Farmers." https://extension.osu.edu/. Accessed February 2026.

Note: Equipment specifications, pricing, and software capabilities referenced in this chapter are accurate as of February 2026. The drone and precision agriculture markets evolve rapidly; verify current options before making purchasing decisions.

Chapter 7: The Multispectral Investment Decision—When the Numbers Work and When They Don't

This chapter isn't here to sell you a multispectral camera. It's here to help you decide—with real numbers and honest math—whether that investment makes sense for your operation right now, or whether your money is better spent somewhere else.

If you've made it through Chapters 4, 5, and 6, you now understand the science of plant reflectance, the vegetation indices that turn light data into management information, and the surprisingly large amount of useful work you can accomplish with a standard RGB camera. The question that remains is whether multispectral capability is worth the additional investment for your particular operation.

This is a question that doesn't have a single right answer. It depends on how many acres you farm, what crops you grow, how variable your fields are, whether you have the equipment to act on variable-rate prescriptions, and—frankly—how much time you're willing to invest in learning a more complex technology. A 2,000-acre corn and soybean operation with variable soils and VRT equipment is in a fundamentally different position than a 300-acre operation with uniform fields and a fixed-rate applicator.

This chapter will walk you through the multispectral basics in plain language, lay out the scenarios where the investment clearly makes sense, be equally clear about when it doesn't, run the break-even math so you can plug in your own numbers, and explore alternative paths that get you multispectral data without the full capital commitment. The goal is a decision you can make with confidence—not one driven by a salesperson's enthusiasm or a neighbor's new toy.

Multispectral Basics: What You're Actually Buying

What It Is

A multispectral camera is a camera system with multiple individual sensors, each designed to capture light at a specific wavelength band. A typical agricultural multispectral camera includes five sensors: one each for blue, green, red, red edge, and near-infrared. Some systems add a sixth band or a dedicated RGB camera for standard color imagery.

Each sensor captures its own image simultaneously. The processing software then stacks these images together, pixel by pixel, so that every point in your field has five or six separate reflectance values associated with it. This stack of data is what allows the software to calculate the vegetation indices we discussed in Chapter 5 – NDVI, NDRE, GNDVI, SAVI, Chlorophyll Index – with true, calibrated accuracy.

Think of it this way: an RGB camera takes a photograph. A multispectral camera takes five photographs at once, each showing a different "color" of light—including colors your eyes can't see. The combination gives you information about the plant that no single photograph could provide.

Why It Costs More

Multispectral cameras are more expensive than standard cameras for several reasons, and understanding these helps explain why the price gap is unlikely to close dramatically anytime soon.

Specialized sensors. Each band requires a dedicated image sensor with precise optical filters that isolate a narrow wavelength range. Manufacturing these to agricultural-grade accuracy is more complex and costly than producing a standard RGB sensor.

Calibration requirements. Multispectral data is only useful if it's calibrated—meaning the values represent actual reflectance percentages, not just relative brightness. Most systems require a calibration panel (a panel with known reflectance values that you photograph before each flight) and an onboard light sensor that compensates for changing illumination. This adds hardware, manufacturing cost, and operational complexity.

Processing complexity. Raw multispectral data requires more sophisticated software to process than RGB imagery. The software must align five separate image sets, apply radiometric corrections, generate calibrated index maps, and output files in formats compatible with precision agriculture equipment. This specialized software carries its own cost, typically through annual subscriptions.

Smaller market. Far fewer people buy multispectral cameras than standard RGB drones. The smaller production volume means manufacturers can't achieve the same economies of scale that drive down consumer drone prices. A DJI Mini sells millions of units; a DJI Mavic 3 Multispectral sells in the tens of thousands.

What's Available in 2026

The multispectral drone market has consolidated around a few key options. Here's what you'll encounter when shopping:

System	Approx. Price	Bands	Key Features
DJI Mavic 3 Multispectral	$4,959+	4 MS + 1 RGB	Integrated system; 43-min flight time; most popular ag drone worldwide
DJI Matrice 350 RTK + MS payload	$10,000–$15,000	5 MS + 1 RGB	Enterprise platform; RTK precision; heavier, more complex
Sentera 6X	$5,500–$7,500	5 MS + 1 RGB	Sensor retrofits to DJI platforms; flexible configuration

System	Approx. Price	Bands	Key Features
MicaSense RedEdge-P	$5,500–$6,500	5 MS	Standalone sensor; mounts on various drones; research-grade
Autel EVO II Pro Enterprise	$7,000–$9,000	Variable	Chinese-manufactured; same FCC Covered List restrictions as DJI

For most farmers considering a first multispectral purchase, the DJI Mavic 3 Multispectral is the default starting point. It's the most widely supported by processing software, has the largest user community for troubleshooting, and offers the best combination of capability and ease of use at its price point. The same DJI regulatory situation from Chapter 3 applies: the Mavic 3M was FCC-authorized prior to December 2025 and remains legal to purchase and fly. Monitor developments as the regulatory landscape continues to evolve.

When Multispectral Makes Sense

Let's be specific about the scenarios where multispectral investment has a clear, defensible return. These aren't the only situations where it can help—but they're the ones where the math works most convincingly.

Large Operations Where Input Savings Scale

The single biggest financial case for multispectral is variable-rate nitrogen management on corn. If you're managing 800 or more acres of corn, even modest per-acre savings add up quickly.

Consider the math. A typical Western Ohio cornfield receives 160–200 pounds of nitrogen per acre. At current prices of roughly $0.45–$0.55 per pound of actual N, that's $72–$110 per acre in nitrogen cost alone. University research consistently shows that precision nitrogen management—guided by NDRE data—can reduce nitrogen use by 10–20%

while maintaining or improving yields. Even at a conservative 10% savings:

NITROGEN SAVINGS MATH

- *800 acres of corn x 180 lbs N/acre x 10% savings = 14,400 lbs N saved*
- *14,400 lbs x $0.50/lb = $7,200 in annual nitrogen savings*
- *That's $7,200 in Year 1 against a $5,000–$6,000 multispectral investment. Payback in under 12 months — and the drone is still flying in Year 2.*

This calculation is conservative. It assumes only a 10% reduction, only on nitrogen, and only on corn acres. Many operations also see savings on fungicide (targeted application rather than blanket spraying), seed (adjusting populations in low-productivity zones), and reduced yield loss from earlier problem detection. The total return is frequently double the nitrogen savings alone.

Fields With High Variability

If your fields are uniform—consistent soil type, flat terrain, even drainage—the benefit of zone-based management is limited. But if your fields look like most of Western Ohio—transitioning between soil types, varying in organic matter, mixing well-drained knobs with poorly drained low spots—variability is your reality, and managing it uniformly costs you money.

Multispectral imagery quantifies that variability with precision. Instead of guessing where the soil transition occurs or estimating how many acres are in the wet zone, you get a calibrated map that shows exactly where performance zones begin and end. If you're already creating management zones from yield maps or soil sampling, multispectral NDRE adds a powerful in-season layer that captures conditions yield maps can't—because yield maps are backward-looking, while NDRE maps are current.

Operations Already Using Variable-Rate Technology

This is a critical qualification. Multispectral data is most valuable when you can act on it with precision. If you own or lease VRT-capable equipment – a variable-rate fertilizer applicator, a planter with variable population capability, or work with a custom applicator who offers VRT services – multispectral maps translate directly into better prescriptions.

Without VRT equipment, multispectral data still has value for scouting and understanding your fields. But the financial return is much harder to capture. Knowing that the northwest 40 acres needs 20 more pounds of nitrogen doesn't help if your applicator can only apply a single rate across the entire field. You're paying for information you can't fully use.

Custom Service Providers and Agronomists

If you're offering—or planning to offer—drone-based crop monitoring services to other farmers, multispectral is essential. Your customers will expect calibrated NDVI and NDRE maps, not RGB approximations. The investment is a cost of doing business, and it's recoverable through service fees. At $8–$15 per acre, flying 500 acres for a single customer generates $4,000–$7,500 in revenue—well on the way to covering your equipment cost in one season.

The same logic applies if you're an independent agronomist incorporating drone data into your consulting practice. Multispectral maps add tangible value to your recommendations and differentiate your services from competitors who rely solely on ground scouting.

When to Skip It (For Now)

Being honest about when multispectral doesn't make sense is just as important as knowing when it does. There's no shame in deciding the investment isn't right for you today. Smart farmers don't buy equipment they can't justify—that discipline is probably why they're still farming.

Operations Under 500 Acres

On a smaller operation, the per-acre savings from precision management may not exceed the annual cost of owning and operating a multispectral system. The equipment costs the same whether you're flying 300 acres or 3,000. The savings scale with acreage; the costs don't.

This doesn't mean smaller operations can't benefit from multispectral data—just that ownership isn't the right path. Hiring a service provider for two or three flights per season is almost always more economical at this scale. We'll cover this option in detail later in this chapter.

Relatively Uniform Fields

If your fields are flat, well-drained, consistent in soil type, and you've been happy with uniform application rates—your yield maps don't show dramatic within-field variability—the case for multispectral is weaker. Precision management tools deliver the biggest returns where variability is greatest. On a field that's already performing consistently, the data may confirm what you already know without revealing actionable new information.

No Variable-Rate Equipment

This is the most common reason to wait. If you can't apply nitrogen at variable rates, the most valuable application of multispectral data—zone-based prescriptions—is off the table. You can still use the data for scouting and monitoring, but at that point, the cost-benefit gap between multispectral and RGB narrows considerably.

If you're planning to add VRT capability in the next year or two, it might make sense to start collecting multispectral data now to build a baseline. But if VRT isn't on your horizon, stick with RGB and invest the $5,000 difference somewhere it'll earn a better return—like that tile project you've been putting off.

First-Year Drone Users

Even if your operation is large enough and variable enough to justify multispectral, I'd still recommend starting with an RGB drone for your first season. Here's why:

There's a learning curve to drone operations—flight planning, image capture, software processing, map interpretation. Adding the additional complexity of multispectral calibration, radiometric correction, and five-band data processing on top of that curve makes the first season overwhelming. Mistakes are more expensive with a $5,000 camera than with a $1,200 one. And the basics of flight operations, weather assessment, and map interpretation apply regardless of camera type.

Spend Season 1 with RGB. Get comfortable. Learn to fly efficiently, process images, and interpret maps. Identify the specific gaps in your data that multispectral would fill. Then invest in multispectral for Season 2 with clear eyes and realistic expectations. You'll get more value from the equipment because you'll know what you're looking for.

When You're Not Ready to Act on the Data

This one is philosophical but practical: why collect data you won't use? If you're not going to change your nitrogen rate based on an NDRE map, if you're not going to adjust your scouting route based on a stress map, if the maps are going to sit in a folder on your computer while you farm the same way you always have—the investment isn't justified.

Data without action is just overhead. The value of multispectral lies entirely in the management decisions it enables. If you're ready to use the data, invest. If you're not, wait until you are.

The Break-Even Math: How Many Acres Justify the Investment?

Let's work through the numbers so you can plug in your own operation and see where you land. We'll use conservative assumptions—if the math works with conservative numbers, you can feel confident in the decision.

Assumptions

Variable	Value	Notes
Multispectral System	$6,000	Mavic 3M + extras, amortized over four years = $1,500/yr
Software Subscription	$1,650/year	Pix4Dfields (annual subscription)
Operating Costs	$400/year	Batteries, insurance, calibration panel replacement
Total Annual Cost of Ownership	$3,550/year	Equipment amortization + software + operating
Nitrogen Price	$0.50/lb actual N	Approximate 2025–2026 pricing
Baseline N Rate	180 lbs/acre	Typical for Western Ohio corn
N Savings from VRT	10%	Conservative; research shows 10–20% possible
Per-Acre N Savings	$9.00/acre	180 lbs x 10% x $0.50

Break-Even Calculation

With an annual cost of ownership of $2,400 and per-acre savings of $9.00 on corn acres:

BREAK-EVEN ACREAGE

$3,550 annual cost ÷ $9.00 savings per acre = approximately 394 corn acres

If you fly and apply variable-rate nitrogen on at least 394 acres of corn, the multispectral system pays for itself through nitrogen savings alone – not counting any additional benefits from scouting efficiency, yield improvement, or fungicide targeting.

Now let's look at how the math scales across different operation sizes:

Corn Acres	Annual N Savings (10%)	Annual System Cost	Net Annual Benefit	Verdict
200 acres	$1,800	$3,550	–$1,750	Doesn't pencil out from nitrogen savings alone
300 acres	$2,700	$3,550	–$850	Below break-even; other benefits may tip it
500 acres	$4,500	$3,550	+$950	Clearly positive; solid investment
800 acres	$7,200	$3,550	+$3,650	Strong return; payback under one year
1,200 acres	$10,800	$3,550	+$7,250	Excellent ROI; no-brainer
2,000 acres	$18,000	$3,550	+$14,450	Exceptional return on investment

A few important notes on this analysis. First, it only counts nitrogen savings on corn. If you also use multispectral data for fungicide targeting, soybean management, or other inputs, the per-acre benefit increases and the break-even acreage drops. Second, the 10% N savings assumption is conservative—published research often shows 15–20%. At 15% savings, the break-even drops to about 263 corn acres. Third, the analysis doesn't include the value of improved yields from earlier problem detection, which can add $5–$15 per acre in avoided losses.

Fourth—and this is the caveat that matters most—these savings only materialize if you actually create and implement variable-rate prescriptions. The drone generates data. The prescription file generates savings. Without that connection, the math doesn't work.

The Alternative Path: Multispectral Data Without the Investment

If the break-even math doesn't work for ownership—or if you're not ready for the full investment—several alternative paths give you access to multispectral data at lower cost and lower commitment.

Hire a Service Provider

This is the most straightforward alternative. A growing number of agricultural service providers, agronomists, and precision agriculture companies offer drone-based multispectral flights. You schedule two or three flights per season at the critical growth stages—typically V6–V8 (pre-side-dress) and VT–R2 (mid-season health check)—and receive processed maps and, in many cases, ready-to-use prescription files.

Service Level	Typical Cost	What You Get
Basic Flight + Maps	$5–$8/acre	Orthomosaic, NDVI map, NDRE map (PDF or digital)
Full Analysis + Prescriptions	$10–$15/acre	Everything above plus zone maps, VRT prescription files, agronomic recommendations
Seasonal Package (3–4 flights)	$12–$20/acre/season	Multiple flights, season-long tracking, comparison maps, end-of-season report

-

For a 500-acre operation, a seasonal package of three flights at $15 per acre runs $7,500. That's more than the annual cost of owning a multispectral drone ($2,400 in our earlier calculation)—which is exactly why ownership makes sense at larger scales. But the service approach requires zero capital investment, zero learning curve for processing, and zero risk of equipment failure. You're buying a complete solution, not a tool.

Finding providers in the Ohio, Indiana, and Illinois region is getting easier. Start with your county agricultural extension office, your co-op's precision agriculture department, or your independent agronomist. Many crop consultants have added drone services in the past few years. The directory in Appendix D lists regional providers.

Free Satellite Imagery

This one surprises most farmers: you can access multispectral satellite imagery of your fields for free, right now, from your computer or phone. It's lower resolution than drone imagery, but it covers every acre you farm at no cost.

The primary source is the European Space Agency's Sentinel-2 satellite constellation, which captures multispectral imagery of the entire planet every five days at a resolution of about 10 meters per pixel (roughly 33 feet). That's not sharp enough to count plants, but it's plenty to identify large-scale stress patterns, track seasonal vigor changes, and compare field performance over time.

Several platforms make Sentinel-2 data accessible without any GIS expertise:

EOS Data Analytics (eosda.com). Free crop monitoring tool that shows NDVI, NDRE, and other indices from satellite data. Draw your field boundaries, and it automatically updates with every new satellite pass. Includes historical data for multi-year comparisons.

Copernicus Browser (dataspace.copernicus.eu). Direct access to Sentinel-2 imagery with a simple map interface. You can view individual bands, generate index maps, and download raw data. Free account required.

Google Earth Engine (earthengine.google.com). More technical, but extremely powerful. Access to decades of satellite imagery. Useful if you want to analyze long-term trends in field performance. Free for noncommercial use.

Farmlens, OneSoil, and similar ag platforms. Several agriculture-specific apps pull satellite data and present it in farmer-friendly formats. Most offer a free tier with basic NDVI monitoring and charge for premium features like prescription generation.

SATELLITE VS. DRONE: THE TRADE-OFF

Resolution: Satellite gives you 33-foot pixels. A drone at 250 feet gives you sub-inch pixels. For scouting-level pattern detection, satellite works. For plant-level analysis and precise prescription maps, you need the drone.

Timing: Satellite passes are on a fixed schedule (every five days for Sentinel-2) and are blocked by cloud cover. You fly your drone when you need to, weather permitting. In a Midwest growing season with frequent cloud cover, you might miss critical windows with satellite alone.

Cost: Satellite is free. You can't beat free. Use it as a complementary data source even if you're flying your own drone.

Partner With Neighbors

A multispectral drone doesn't care who owns the fields it flies over. If a $6,000 investment doesn't pencil out for your 400-acre operation alone, consider pooling resources with one or two neighbors. Three farms totaling 1,200 acres can justify the investment easily, share the operating costs, and take turns flying each other's fields.

This works best when the participating farms are geographically close (minimizing travel time between fields), have similar crop mixes and timing needs, and the group can agree on scheduling priority during critical windows. A simple written agreement covering equipment ownership, maintenance responsibility, scheduling, and liability is worth the time to draft.

Some Farm Bureau chapters and co-ops have facilitated shared equipment arrangements for other precision agriculture tools. The model is well-established in the Midwest. There's no reason it can't extend to drones.

The Stepping-Stone Approach

Here's the path we recommend for most farmers who aren't sure whether multispectral is right for them:

Year	Action	Investment	What You Learn
Year 1	Buy RGB drone; fly your fields; learn the basics	$1,500	Flight operations, map interpretation, where your fields vary
Year 1–2	Use free satellite NDVI to monitor seasonal patterns	$0	Which fields have the most variability; when stress appears
Year 2	Hire a multispectral service for 2–3 flights	$2,000–$5,000	Whether NDRE data changes your management decisions
Year 3	Decide: buy multispectral, continue hiring, or stay RGB	Varies	You know exactly what you need, based on two seasons of real experience

This approach costs more in total over three years than buying multispectral upfront. But it dramatically reduces the risk of a wrong decision. You're investing incrementally, learning at each step, and making the big purchase only after you've confirmed that the data actually changes how you farm. That's a smarter use of capital than betting $6,000 on a technology you haven't proven on your own fields.

A Note on Processing Software

Multispectral data requires specialized software that RGB data doesn't. The processing pipeline is more complex: aligning multiple bands, applying radiometric calibration, correcting for sun angle and illumination changes, and generating calibrated index maps. Your free DJI Fly app won't handle this.

When budgeting for a multispectral system, factor in software as an ongoing cost. Here are the main options:

Software	Annual Cost	Best For
Pix4Dfields	$1,650/year	Best agriculture-specific value; easy to learn; runs locally on your computer
DroneDeploy	$1,908–$5,988/year	Cloud-based; powerful but expensive; best for service providers flying many farms
OpenDroneMap (WebODM)	Free	Open-source; steeper learning curve; runs locally; surprisingly capable for free software
Sentera FieldAgent	Varies	Integrates with Sentera sensors; good if you're in the Sentera ecosystem
Solvi	$600–$1,200/year	Cloud-based; agriculture-focused; good balance of capability and price

For most farmers processing their own multispectral data, Pix4Dfields at $1,650 per year offers the best combination of capability, ease of use, and value. If you're willing to invest time in learning, OpenDroneMap is free and handles multispectral data competently. DroneDeploy is excellent but priced for service providers and large operations, not individual farmers.

The Decision Framework: A Summary

To pull everything together, here's a decision framework you can use to determine your best path forward.

BUY MULTISPECTRAL IF...

- *You farm 500+ acres of corn (break-even threshold with conservative*
- *assumptions)*
- *You own or have access to variable-rate application equipment*
- *Your fields have significant within-field variability*
- *You've already flown RGB for at least one season and understand*
- *the basics*
- *You're committed to using the data to change management decisions*
- *You're offering drone services to other farmers*

HIRE A SERVICE INSTEAD IF...

- *You farm 200–500 acres and want multispectral data without the capital investment*
- *You want to test whether multispectral data changes your decisions before committing*
- *You don't have time to learn multispectral processing and calibration*
- *You need two or three flights per season, not weekly access*

STICK WITH RGB (FOR NOW) IF...

- *You're in your first year of drone use and still learning the basics*
- *You farm under 200 acres with relatively uniform fields*
- *You don't have variable-rate equipment and aren't planning to add it soon*
- *Your primary needs are scouting, stand counts, drainage mapping, and documentation*
- *The $5,000–$6,000 investment would strain your operating budget*

Whatever path you choose, remember that this technology is a tool, not a destination. The destination is better management decisions that improve your bottom line. Whether you get there with a $1,200 RGB drone, a hired multispectral service, or a $6,000 camera system of your own—the value is in what you do with the information, not in the equipment that generates it.

Chapter Summary

- A multispectral camera captures light in five or more wavelength bands simultaneously, enabling true NDVI, NDRE, and other calibrated vegetation indices. Typical cost is $5,000–$8,000 for a complete system.
- Multispectral investment makes the strongest financial case on operations of 500+ corn acres with variable fields and variable-rate application equipment. Conservative break-even is approximately 394 corn acres based on 10% nitrogen savings alone, accounting for current software costs.
- Skip multispectral (for now) if you farm under 200 uniform acres, lack VRT equipment, are new to drones, or aren't ready to change management practices based on the data.
- The break-even math scales dramatically with acreage: a 500-acre corn operation can expect $2,100 in net annual benefit from nitrogen savings alone; at 1,200 acres, that grows to $8,400.
- Alternative paths include hiring a service provider ($5–$15 per acre), using free satellite imagery from Sentinel-2, partnering with neighbors to share equipment costs, or following a stepping-stone approach from RGB to hired services to ownership.
- Processing software is an ongoing cost that must be factored into the multispectral budget. Pix4Dfields ($1,650/year) offers the best value for individual farmers who need full multispectral processing; OpenDroneMap is free but has a steeper learning curve.

- The stepping-stone approach—RGB in Year 1, hired multispectral flights in Year 2, ownership decision in Year 3—reduces risk and ensures you invest based on experience rather than marketing.
- The value of any drone system lies entirely in the management decisions it enables. Data without action is just overhead.

Chapter 7 Notes and References

1. DJI. "DJI Mavic 3 Multispectral." https://enterprise.dji.com/mavic-3-m. Accessed February 2026.

2. Sentera. "6X Multispectral Sensor." https://sentera.com/6x. Accessed February 2026.

3. MicaSense. "RedEdge-P Multispectral Sensor." https://micasense.com/rededge-p. Accessed February 2026.

4. Thompson, L. and Puntel, L. "Optimize your Nitrogen Investment with Precision Agriculture Technologies." University of Nebraska–Lincoln CropWatch, November 2021. https://cropwatch.unl.edu/2021/optimize-your-nitrogen-investment-precision-agriculture-technologies/

5. Pix4D. "Pix4Dfields: Agriculture Drone Mapping Software." https://www.pix4d.com/product/pix4dfields. Accessed February 2026.

6. European Space Agency. "Sentinel-2 Mission." https://sentinel.esa.int/web/sentinel/missions/sentinel-2. Accessed February 2026.

7. EOS Data Analytics. "Crop Monitoring." https://eos.com/products/crop-monitoring. Accessed February 2026.

8. OpenDroneMap. "WebODM: Open-Source Drone Mapping." https://www.opendronemap.org/webodm. Accessed February 2026.

9. U.S. Government Accountability Office. "Precision Agriculture: Benefits and Challenges for Technology Adoption and Use." GAO-24-105962, January 2024. https://www.gao.gov/products/gao-24-105962

10. USDA Economic Research Service. "Precision Agriculture in the Digital Age." Economic Information Bulletin No. 248, February 2023. https://www.ers.usda.gov/publications/pub-details?pubid=105893

Note: Equipment pricing, software costs, and nitrogen prices referenced in this chapter are approximate and based on conditions as of early 2026. Input costs and equipment pricing fluctuate; verify current numbers when making your own calculations. This chapter provides a framework for analysis, not a guarantee of specific financial returns.

PART THREE — Flight Operations – From Truck to Sky and Back

Chapters 8–11

You understand the science. Now it's time to fly. Part III covers everything from pre-flight planning to processed maps to your first full season of drone-assisted management. Chapter 8 addresses flight planning for real-world conditions—altitude, overlap, weather, and battery management. Chapter 9 walks you through turning raw images into usable maps with processing software. Chapter 10 teaches you to read those maps like a farmer—ground-truthing anomalies and connecting what you see on screen to what's happening in the field. Chapter 11 is your 12-week action plan: a structured, phase-by-phase schedule for your first season that takes you from unboxing to confidence.

Chapter 8: Flight Planning for the Real World

You don't need to be a pilot. You don't need a detailed flight plan filed with anyone. What you need is thirty minutes, a charged battery, and a basic understanding of when, where, and how high to fly. This chapter gives you all three.

Parts One and Two gave you the "why"—why drone technology can benefit your operation, what the science is behind plant reflectance, and how to decide which equipment makes sense. Now we get to the "how." Part Three is about operations: the practical, boots-on-the-ground (and drone-in-the-sky) reality of flying your fields and turning those flights into useful information.

This chapter focuses on flight planning—the decisions you make before and during a flight that determine whether you come home with useful data or a memory card full of unusable images. We'll cover timing (when to fly during the season and during the day), altitude and overlap settings (the two decisions that most affect data quality), how much ground you can realistically cover per battery, and a simple 30-minute routine that fits into the rhythm of a working farm.

The goal isn't to turn you into a drone operations specialist. It's to give you a repeatable system that produces reliable results without eating up your day. Most of the farmers we've worked with find that after three or four flights, the process becomes second nature—about as complicated as hooking up a grain cart.

When to Fly: The Seasonal Calendar

The most common question from farmers new to drone monitoring is "How often should I fly?" The answer isn't "as often as possible." More flights don't automatically mean more value. What matters is flying at the right growth stages—the moments when aerial information can actually change a management decision.

For Midwest row crops, most of the actionable value comes from three to four strategic flights per season, timed to specific growth stages. Here's the schedule, broken down by crop.

Corn Flight Schedule

Flight	Growth Stage	Timing (Approx.)	Primary Camera Needed	Purpose
Flight 1	V4–V6	Late May / Early June	RGB is sufficient	Stand assessment, emergence uniformity, early stress, planter performance
Flight 2	V10–VT	Late June / Early July	Multispectral preferred	Nitrogen status (NDRE), pre-tassel health baseline, equipment issue confirmation
Flight 3	R2–R3	Late July / Early August	Multispectral preferred	Pollination stress, disease onset (tar spot, gray leaf spot), canopy health
Flight 4 (optional)	R5–R6	Late August / September	Either	Late-season stress, yield prediction, harvest planning

>

>

Soybean Flight Schedule

Flight	Growth Stage	Timing (Approx.)	Primary Camera Needed	Purpose
Flight 1	V3–V4	Late May / Early June	RGB is sufficient	Stand assessment, early stress, weed pressure before canopy closure
Flight 2	R2–R3	Mid-July	Multispectral preferred	Canopy closure check, early disease (frogeye, white mold risk areas)
Flight 3	R5–R6	August	Either	Late-season disease, maturity assessment, stress patterns

>

>

Wheat Flight Schedule

Flight	Growth Stage	Timing (Approx.)	Primary Camera Needed	Purpose
Flight 1	Spring greenup	March / Early April	RGB is sufficient	Winter kill assessment, stand thickness, early vigor
Flight 2	Flag leaf – heading	Late April / May	Multispectral preferred	Disease pressure (rust, Septoria), nitrogen status
Flight 3	Grain fill	Late May / June	Either	Late disease, harvest timing, stress assessment

THE PRACTICAL SUMMARY

For a typical corn/soybean rotation in Western Ohio or Eastern Indiana, plan on three to four flights per field during the growing season. That's roughly one flight every three to four weeks from late May through August.

Total time commitment for a 500-acre operation: approximately 15–20 hours across the entire season, including setup, flying, and battery swaps.

Timing Beyond Growth Stage: Special Situations

The seasonal schedule covers routine monitoring, but some of the most valuable flights happen outside the regular calendar—triggered by events rather than growth stages.

After significant rain events. Fly within 24–48 hours of a heavy rain (1.5 inches or more) to map drainage patterns, identify tile failures, and document ponding. RGB is perfect for this. These opportunistic flights often deliver more immediate, actionable information than any scheduled flight.

After storm damage. Hail, straight-line winds, or flooding – fly as soon as conditions allow. The imagery provides documentation for insurance claims and helps you make replant or abandonment decisions based on data rather than guesswork.

Before side-dress applications. If you're using multispectral data for nitrogen prescriptions, time your flight 3–5 days before the applicator is scheduled. This gives you time to process the data, generate the prescription file, and load it onto the equipment.

When something looks "off" from the road. You're driving past a field and notice a color change or a patch that doesn't look right. Instead of walking in from the edge and trying to assess the extent from ground level, fly the field. Five minutes in the air can tell you whether it's a 2-acre problem or a 20-acre problem – and where exactly the boundaries are.

When to Fly: Time of Day and Weather

The seasonal calendar tells you which week to fly. The time of day and weather tell you which hour. Getting these right is the difference between clean, usable data and maps that mislead you.

Best Conditions

Overcast skies are ideal. This surprises many people. A bright, sunny day feels like perfect drone weather, but overcast conditions produce the most uniform, reliable imagery. Diffuse light from an overcast sky eliminates shadows, reduces glare, and creates even illumination across the entire field. Your maps will have consistent color and reflectance values from edge to edge.

If flying on sunny days, stay within two hours of solar noon. When the sun is directly overhead (roughly 11:00 AM to 1:00 PM in the Midwest during summer), shadows are shortest and most uniform. Early morning or late afternoon flights cast long shadows that create false variability in your maps – darker strips on one side of every row that look like stress but are just shade.

Wind under 15 mph. The DJI Mini series can handle moderate wind, but image quality degrades as the drone works harder to hold position. At 15+ mph, you'll see blurred images, inconsistent overlap, and the drone may drift off its planned flight path. Below 10 mph is ideal. Above 20 mph, stay on the ground.

Dry canopy. Dew, recent rain, or irrigation water on the leaves changes reflectance values, particularly in the near-infrared bands. For the most reliable data, wait at least two hours after the canopy dries. Early morning dew flights can produce misleading multispectral data, even though the RGB imagery looks fine.

No rain, no fog. This should be obvious, but it's worth stating: don't fly in rain, drizzle, or fog. The Mini series drones are not waterproof, and moisture on the camera lens ruins imagery. Light mist might not damage the drone, but it will destroy your data quality.

The Partly Cloudy Problem

Partly cloudy days are the trickiest conditions for aerial monitoring, and they're extremely common during a Midwest summer. The issue is simple: when a cloud shadow passes over part of your field during a flight, the images captured in shadow have dramatically different brightness than images captured in sun. When the software stitches these together, you get a map with artificial light and dark patches that have nothing to do with crop health.

If you must fly on a partly cloudy day, watch the sky and try to fly when the sun is consistently behind a cloud bank or consistently clear. A full cloud cover and a clear sky both work – it's the mix that causes problems. If you notice a shadow crossing the field during your flight, the best option is to restart that section when conditions stabilize.

For multispectral flights, the partly cloudy problem is worse than for RGB. Multispectral sensors rely on consistent illumination for calibrated reflectance values. If you're generating NDRE maps for nitrogen prescriptions, the accuracy of those prescriptions depends on consistent lighting. Schedule your critical multispectral flights for overcast or clear days if at all possible.

WEATHER DECISION GUIDE

Overcast, calm: Ideal. Fly any time of day. Best conditions for multispectral.

Clear and sunny: Good. Fly within two hours of solar noon to minimize shadows.

Partly cloudy: Caution. Wait for consistent conditions. Avoid for multispectral.

Wind 10–15 mph: Acceptable. Expect slightly reduced image quality. Monitor battery drain.

Wind 15–20 mph: Marginal. Fly only if necessary. Reduce altitude slightly for stability.

Wind 20+ mph: No. Stay on the ground. Not worth the risk to equipment or data quality.

Rain, fog, drizzle: No. Never fly in precipitation.

Flight Height and Overlap: The Two Settings That Matter Most

Once you've decided when to fly, the next two decisions—altitude and image overlap—determine the quality and usefulness of your data. Get these right, and your software will produce clean, accurate maps. Get them wrong, and you'll end up with gaps, distortions, or maps too blurry to be useful.

Flight Altitude

The sweet spot for crop monitoring: 200–300 feet AGL. AGL means "above ground level"—the height above the terrain directly below you, not above sea level. At this altitude range, you get a resolution of roughly 1–3 inches per pixel with a standard RGB camera—more than enough to see individual plants, count rows, and identify color differences.

Why not fly lower? You can, and for some applications – like detailed stand counts at V2 – dropping to 100–150 feet gives you even sharper images. But lower altitude means less ground covered per image, which means more images needed, more overlap required, longer flights, and more battery consumption. For routine monitoring, it's not worth the trade-off.

Why not fly higher? Above 300 feet, resolution drops off. Individual plants become harder to distinguish. Color differences become subtler. And you're getting closer to the 400-foot legal ceiling without gaining much efficiency – the increase in coverage per image doesn't offset the loss of detail.

ALTITUDE AND COVERAGE REFERENCE

Altitude (AGL)	Approx. Resolution	Coverage per Image	Best For
100–150 ft	0.5–1 inch/pixel	~1.5 acres	Detailed stand counts, plant-level assessment, small problem areas
200–250 ft	1–2 inches/pixel	~3–4 acres	General crop monitoring, health assessment, weed scouting (best all-around)

Altitude (AGL)	Approx. Resolution	Coverage per Image	Best For
250–300 ft	2–3 inches/pixel	~5–6 acres	Broad field overview, large-scale pattern detection, drainage mapping

>

>

Our recommendation: fly at 250 feet for most routine monitoring.

It's a reliable all-purpose altitude that balances resolution, coverage, and battery efficiency. Drop to 150 feet for detailed stand counts. Go up to 300 feet if you're flying large fields and just need a general overview. Consistency matters more than optimization—if you fly the same field at the same altitude every time, your maps will be directly comparable across dates.

Image Overlap

Image overlap is the percentage of each photo that overlaps with the adjacent photo. Your mapping software uses these overlapping areas to stitch images together into a seamless map. Without sufficient overlap, the software can't find enough common reference points between images, and your map will have gaps, warps, or holes.

There are two types of overlap:

Front overlap (forward overlap) is the overlap between consecutive images along the direction of flight. The drone takes photos in sequence as it flies a line, and each photo shares a percentage with the one before and after it.

Side overlap (lateral overlap) is the overlap between adjacent flight lines – the strips the drone flies back and forth across the field. Each pass shares a percentage of coverage with the pass next to it.

RECOMMENDED OVERLAP SETTINGS

For crop monitoring with automated flight apps:

Front overlap: 75%

Side overlap: 70%

These settings provide enough redundancy for reliable stitching while keeping flight times reasonable. Some software defaults to 80% / 70% or 80% / 80%—that works too, but your flights will take longer and produce more images to process.

If you're getting gaps or stitching errors in your maps, increase overlap by 5%. If your flights are taking too long, decrease by 5%. But don't go below 65% on either axis – the stitching quality degrades rapidly below that threshold.

Why Overlap Matters: A Practical Explanation

Imagine you're laying out aerial photos on your kitchen table, trying to piece together a picture of your field. If each photo shows exactly the same area as the one next to it – 100% overlap – you'd need the most photos but could match them perfectly. If each photo had zero overlap, you'd just be placing non-overlapping tiles, and any slight misalignment would create gaps or double-coverage.

At 75% front overlap, each photo shares three-quarters of its view with the next photo. The software identifies matching features – a distinctive plant, a row pattern, a soil mark – across multiple photos and uses those matches to precisely align everything. The more overlap, the more matching points, and the more accurate the final map.

Over agricultural fields, this matters because crops are relatively uniform. A cornfield at V10 looks like a sea of green – there aren't a lot of distinctive features for the software to grab onto. Higher overlap gives the software more data to work with, which is why agricultural flights typically need more overlap than, say, mapping a construction site with buildings and vehicles as reference points.

Coverage Per Battery: The Real-World Numbers

Drone manufacturers advertise maximum flight times of 34–52 minutes (depending on model and battery). In agricultural operations, real-world flight time is always less. Wind resistance, GPS corrections, automated turns at the end of each pass, and the reserve you should always maintain before landing all eat into the advertised number. Plan on 25–30 minutes of usable flight time per battery.

How much ground that covers depends on your altitude, overlap settings, and the drone's speed. Here's what to expect with the DJI Mini 4 Pro or

Mini 5 Pro at typical agricultural settings:

FLIGHT PLANNING REFERENCE

Altitude (AGL)	Overlap (F/S)	Speed	Acres per Battery	Notes
150 ft	80%/75%	~12 mph	25–35 acres	Detailed stand counts; use for small areas only
200 ft	75%/70%	~15 mph	40–55 acres	
250 ft	75%/70%	~18 mph	55–75 acres	
300 ft	75%/70%	~20 mph	70–90 acres	

>

These numbers assume moderate wind (under 10 mph) and relatively flat terrain. In stronger wind, battery consumption increases and coverage decreases. If your fields have significant terrain variation, the drone works harder to maintain altitude, which also reduces coverage.

Planning Your Battery Budget

Before heading to the field, do a quick calculation: how many batteries will this job take?

BATTERY PLANNING EXAMPLE: QUARTER-SECTION COVERAGE

Field size: 160 acres (quarter section)

Altitude: 250 feet

Expected coverage: ~65 acres per battery

Batteries needed: 160 ÷ 65 = 2.5 -> three batteries

With the Fly More Combo's three batteries, you can cover a quarter section in one trip. For a half section (320 acres), you'll need to recharge or bring extra batteries. This is why we recommended the Fly More Combo in Chapter 2—three batteries is the minimum for practical agricultural use.

A practical tip: always land with at least 20% battery remaining. The drone will warn you at 20% and force an automatic landing at a lower threshold. Landing at 20% gives you a safety margin for unexpected wind gusts or if the return-to-home path is longer than expected. Better to lose two minutes of flight time than to lose a drone in the middle of a cornfield.

Automated Flight Planning: Let the App Do the Work

Here's the good news for anyone intimidated by the idea of manually flying a drone back and forth across a 160-acre field: you don't have to. Automated flight planning apps handle the entire process. You draw the field boundary on a map, set your altitude and overlap, tap "start," and the drone flies a systematic grid pattern on its own. You stand there, watch it work, and swap batteries when needed.

This is not a nice-to-have feature. For agricultural mapping, automated flight is essential. Manual flying can't achieve the consistent altitude, speed, and overlap that mapping software requires. You'll use manual flight for quick visual checks ("let me look at that spot"), but every mapping flight should be automated.

App Options

DJI Fly app (free, built-in). The app that comes with your DJI drone includes a "Hyperlapse" waypoint mode and basic mapping capability. For simple RGB flights, it's adequate. The interface is straightforward: draw your area, set altitude, tap go. Limitations include less control over overlap settings and fewer export options compared to dedicated mapping apps.

DJI Pilot 2 (free, for enterprise drones). If you're flying a Mavic 3 Multispectral or Matrice platform, DJI Pilot 2 includes a dedicated mapping mission mode with full control over altitude, overlap, speed, and camera settings. This is the standard for multispectral agricultural flights on DJI hardware.

Dronelink (free tier + paid). A third-party app that works with DJI drones and offers robust flight planning features. The free tier covers basic mapping missions. The paid tier ($15–$30/month) adds advanced features like terrain following—where the drone automatically adjusts altitude to maintain a consistent height above ground on hilly terrain. Worth considering if your fields have significant elevation changes.

Litchi ($25 one-time). A popular third-party flight app with waypoint missions and automated grid patterns. One-time purchase, no subscription. Compatible with most DJI consumer drones. A solid option for farmers who want more control than the DJI Fly app provides without paying a monthly fee.

DroneDeploy and Pix4Dcapture (subscription). Both include flight planning modules tied to their processing platforms. If you're already paying for their software, the integrated flight planning is seamless – the app plans the flight and the cloud platform processes the images automatically. But the subscription cost is only justified if you're using the full processing pipeline.

OUR RECOMMENDATION FOR BEGINNERS

Start with the free DJI Fly app for your first several flights. Learn the basics of automated missions without adding another piece of software to learn. Once you're comfortable with the process and want more control over mission settings, try Litchi ($25 one-time) or the Dronelink free tier. Move to DroneDeploy or Pix4D only if you're *using their processing software.*

Setting Up a Flight Mission: Step by Step

1. Regardless of which app you use, the basic process is the same. Here's the workflow for a typical mapping flight:
2. Open the app and select "Mapping Mission" (or equivalent – "Area Mission," "Survey," or "Grid" depending on the app).
3. Draw your field boundary. Most apps let you tap corners on a satellite map. You can also import field boundaries from your farm management software or create them in Google Earth. The boundary doesn't need to be perfect – extend it 50–100 feet beyond the field edge to ensure complete coverage.
4. Set your altitude. 250 feet for routine monitoring. 150 feet for stand counts. 300 feet for large-area overview.
5. Set overlap. 75% front, 70% side for most applications. The app will show you the estimated flight time, number of photos, and ground coverage.
6. Choose flight direction. Most apps let you rotate the grid pattern. For agricultural fields, align the flight lines perpendicular to the crop rows if possible. This gives the stitching software more distinctive features to work with (it can see row patterns changing rather than identical rows stretching the length of the field). If that's not practical, any direction works – it's a preference, not a requirement.
7. Review the estimated flight time. If it exceeds one battery, the app will either break the mission into segments automatically or prompt you to swap batteries mid-mission. Either way, know before you launch how many batteries you'll need.
8. Launch and monitor. Tap "Start," and the drone takes over. It climbs to altitude, flies to the first waypoint, and begins the grid. Your job is to watch: maintain visual line of sight, monitor battery level, and be ready to take manual control if anything unexpected happens (another aircraft, sudden weather change, equipment malfunction).

Sample Flight Plans by Field Size

To give you a concrete sense of what's involved, here are flight plan estimates for common field sizes. These assume 250-foot altitude, 75%/70% overlap, moderate wind, and a DJI Mini 4/5 Pro.

Field Size	Flight Time	Batteries	Images	Total Time (with setup)
40 acres	12–15 minutes	1	~200–280	20–25 minutes
80 acres	22–28 minutes	1 (tight) or 2	~400–550	30–40 minutes
160 acres	45–55 minutes	2–3	~800–1,100	60–75 minutes
320 acres	90–110 minutes	4–5	~1,600–2,200	2–2.5 hours
640 acres (section)	180–220 minutes	8–10	~3,200–4,400	4–5 hours

A few things jump out from this table. First, a quarter section—the most common field unit in the Midwest—is a roughly one-hour job from truck arrival to departure. That's manageable. Second, once you get above 320 acres in a single session, you're looking at a significant time commitment. For larger operations, you'll likely fly fields over multiple days rather than marathoning through a full section at once.

Third, notice the image counts. A 160-acre field at these settings produces 800–1,100 images. That's a lot of data. A 128GB microSD card holds approximately 3,000–4,000 high-resolution images—enough for several fields before you need to offload. Carry a spare card or a laptop to transfer files in the field if you're flying multiple fields in one day.

The 30-Minute Flight Routine

One of the biggest barriers to drone adoption isn't cost or technology—it's time. During the growing season, every hour matters. If flying a field feels like a half-day project, it won't happen. The goal is a repeatable routine that fits into your day the way checking cattle or running to the co-op does—a task, not an event.

For a typical 80-acre field, the entire process – arrival to departure – takes about 30 minutes. Here's the breakdown:

Pre-Flight (3–4 Minutes)

Park and set up. Pull off on the field road or headland. Set your landing pad on a flat, clean surface (the tailgate of your truck works fine). Remove the drone from its case, unfold the arms, attach the propellers if they're not already on.

Power on and connect. Turn on the controller, then the drone. Wait for GPS lock—this takes 30–60 seconds. The app will show “Ready to Fly” when the drone has enough satellite connections for stable flight.

Quick visual check. Propellers secure? Camera lens clean? Battery fully charged? microSD card inserted? Airspace clear? Any TFRs? This should take about 30 seconds once it's habit.

Load your mission. Open the flight planning app, select the pre-programmed mission for this field (or create a quick one if it's your first time here), and verify the settings: altitude, overlap, boundary.

Flight (20–25 Minutes)

Launch. Tap "Takeoff" in the app. The drone lifts to a hover at about 5 feet. Then tap "Start Mission." It climbs to altitude and begins the grid pattern.

Monitor. Keep the drone in sight. Glance at the controller screen periodically to check battery level, signal strength, and mission progress. Most of your time is spent standing or sitting by the truck, watching the drone work its pattern.

Battery swap (if needed). When the first battery hits 20–25%, the drone will return to its launch point and land (or the app will prompt you to bring it back). Swap the battery – a 30-second operation once you've done it a few times – and resume the mission. Most apps pick up where they left off.

Post-Flight (2–3 Minutes)

Land and power down. When the mission completes, the drone returns to launch and lands. Power off the drone, then the controller.

Quick equipment check. Inspect the drone for any damage or debris (grass, dust, insects in the motors). Fold the arms, secure the propellers, pack it in the case.

Log the flight. Take 60 seconds to record the date, field, conditions, and any observations in your flight log. This discipline will save you time later when you're trying to remember which flight went with which field. A template is in Appendix H.

THE PRE-FLIGHT CHECKLIST (LAMINATED, TRUCK-READY)

- *[ ] Airspace checked (B4UFLY or Aloft app)*
- *[ ] Weather acceptable (wind < 15 mph, no precipitation, visibility > 3 miles)*
- *[ ] Batteries charged (plan + one spare)*
- *[ ] microSD card inserted with space available*
- *[ ] Controller charged*
- *[ ] Propellers secure, camera lens clean*
- *[ ] GPS lock acquired (green light / app confirmation)*
- *[ ] Flight mission loaded and verified*
- *[ ] Launch area clear of people, vehicles, overhead lines*
- *[ ] Return-to-home altitude set (above obstacles)*

Seasonal Flight Calendar: Western Ohio / Eastern Indiana

Here's a month-by-month calendar for planning your flights across a typical corn and soybean growing season. Dates are approximate and will shift with planting timing and weather.

Month	Corn Activity	Soybean Activity	Other Opportunities
March–April	–	–	Wheat greenup assessment (Flight one). Pre-season drainage check after snowmelt.

Month	Corn Activity	Soybean Activity	Other Opportunities
May (late)	V2–V4: Stand assessment, emergence, planter performance.	V2–V3: Early stand assessment, emergence.	Post-rain drainage flights. Document any replant areas.
June (early)	V4–V6: First routine monitoring flight. Finalize stand counts. Weed scouting.	V4–V5: Weed scouting, identification before canopy closure.	Equipment problem patterns.
June (early)	**V4–V6: First routine monitoring flight. Finalize stand counts. Weed scouting.**	**V4–V5: Weed scouting, identification before canopy closure.**	**Equipment problem patterns.**
June (late)	V8–V10: NDRE flight for side-dress nitrogen decisions. (Flight 2).	R1–R2: Canopy closure. Baseline health assessment.	Wheat heading: Critical disease and nitrogen assessment.
July	VT–R2: Pollination stress, disease onset monitoring.	R3: Disease scouting (frogeye, white mold risk).	Wheat harvest; fall flights if needed.
August	R3–R5: Late-season stress, yield prediction, disease tracking.	R5–R6: Maturity assessment, late disease.	Cover crop planning; identify problem areas for fall work.
September	R6: Final assessment before harvest if needed.	R7–R8: Harvest timing assessment.	Post-harvest drainage flights. Tile installation planning.

September	R6: Final assessment before harvest if needed.	R7–R8: Harvest timing assessment.	Post-harvest drainage flights. Tile installation planning.
October–Nov.	–	–	Post-harvest field assessment. Compaction documentation. Wheat emergence (if fall-planted).

You don't need to fly every window in this calendar. Pick the flights that align with your highest-value decisions—typically the V4–V6 stand assessment and the V8–V10 nitrogen assessment for corn—and add others as time and interest allow. Three flights per field across the season is a solid starting point. You can always add more in subsequent years as you build confidence and identify where additional data would change your management.

Terrain and Special Situations

Hilly or Rolling Fields

The Midwest isn't the Great Plains. Western Ohio and Eastern Indiana have enough terrain variation—creek bottoms, moraines, glacial ridges—that maintaining a consistent altitude above ground requires attention. When you set your drone to fly at 250 feet, it maintains 250 feet above its launch point—not 250 feet above the ground. If the field drops 50 feet from one end to the other, your drone is 250 feet up on one end and 300 feet up on the other.

For most crop monitoring, this variation is acceptable. A 50-foot difference in altitude produces a minor change in resolution that won't affect your management decisions. But if your fields have terrain changes greater than 75–100 feet, consider using an app with terrain-following capability (like Dronelink), which adjusts the drone's altitude throughout the flight to maintain a consistent height above the actual ground surface.

Obstacles and Hazards

Before your first flight over any field, mentally catalog the obstacles. Power lines, cell towers, grain bins, tree lines, windmills—anything that rises above crop height. Your automated flight plan doesn't know about these. At 250 feet, you'll clear most obstacles in an agricultural setting, but power line poles can reach 60–80 feet and cell towers much higher.

The DJI Mini series' obstacle avoidance sensors help, but they're not infallible—they work best with solid objects at close range and may not detect thin wires. The safest approach is to identify obstacles before the flight and either adjust the flight plan to avoid them or set the return-to-home altitude above the tallest obstacle in the area.

Multiple Fields in One Session

If you're flying several fields in sequence, plan your route like you'd plan a day of field work—minimize windshield time. Pre-program missions for each field so you can land at one field, drive to the next, and launch again without fiddling with settings. Carry a cooler with cold batteries (heat degrades lithium-polymer performance) and a spare microSD card.

A realistic pace: two to three fields per hour for 40–80 acre fields, including drive time. For a 500-acre operation split across six or seven fields, you can cover everything in a half-day. The first season takes longer as you set up missions and learn the routine; by the second season, it's mechanical.

Chapter Summary

- Three to four strategic flights per season – timed to key growth stages – deliver most of the value from drone monitoring. More flights don't automatically mean more value.
- For corn, the highest-value flights are at V4–V6 (stand assessment) and V8–V12 (nitrogen status). For soybeans, target V3–V4 (stand

assessment) and R2–R3 (canopy and disease).

- Event-driven flights—after storms, heavy rain, or when something looks off from the road—often deliver the most immediately actionable information.
- Overcast conditions are ideal. Fly within two hours of solar noon on sunny days. Avoid partly cloudy conditions for multispectral flights. Wind under 15 mph.
- Fly at 200–300 feet AGL with 75% front overlap and 70% side overlap. These settings balance image quality, coverage, and battery efficiency for most monitoring applications.
- Expect 55–75 acres per battery at 250 feet with a DJI Mini 4/5 Pro. A quarter section (160 acres) takes 2–3 batteries and about one hour total.
- Use automated flight planning apps – not manual flying – for all mapping missions. Start with the free DJI Fly app and upgrade as needed.
- The entire flight process – setup through pack-up – fits into a 30-minute routine for an 80-acre field. It's a task, not an event.
- Pre-flight checklist, flight log, and seasonal calendar templates are included in the appendices for field use.
- Always land with 20% battery remaining, check airspace before every flight, and catalog obstacles before flying any field for the first time.

Chapter 8 Notes and References

1. DJI. "DJI Mini 4 Pro User Manual." Flight performance and battery specifications. https://store.dji.com/product/dji-mini-4-pro. Accessed February 2026.

2. DJI. "DJI Mini 5 Pro Specifications." https://www.dji.com/mini-5-pro. Accessed February 2026.

3. Pix4D. "How to Verify Image Quality for Photogrammetry." Knowledge Base. https://support.pix4d.com/. Accessed February 2026.

4. Federal Aviation Administration. "14 CFR Part 107 – Small Unmanned Aircraft Systems." Flight altitude and operating regulations.

https://www.ecfr.gov/current/title-14/chapter-I/subchapter-F/part-107

5. Torres-Sánchez, J., López-Granados, F., and Peña, J.M. "An automatic object-based method for optimal thresholding in drone images." Computers and Electronics in Agriculture, 114, 43–52, 2015. Research on optimal flight parameters for agricultural drone imagery.

6. Mesas-Carrascosa, F.J., Torres-Sánchez, J., Clavero-Rumbao, I., et al. "Assessing Optimal Flight Parameters for Generating Accurate Multispectral Orthomosaics by drone." Remote Sensing, 7(10), 12793–12814, 2015.

7. Dronelink. "Terrain Following for Agricultural Mapping." https://www.dronelink.com/. Accessed February 2026.

8. Litchi. "Mission Planning for DJI Drones." https://flylitchi.com/. Accessed February 2026.

9. Ohio State University Extension. "Corn Growth Stages and Development." Agronomy Guide. https://extension.osu.edu/. Accessed February 2026.

10. Purdue University Extension. "Soybean Growth and Development." https://ag.purdue.edu/. Accessed February 2026.

Note: Flight times, coverage estimates, and battery performance

referenced in this chapter are based on real-world agricultural flight experience with DJI Mini series drones in Midwest conditions. Actual performance will vary with wind, temperature, terrain, and specific equipment configuration. Use these estimates for planning purposes and adjust based on your own experience.

Chapter 9: Turning Flights Into Maps – Software That Won't Break the Bank

Your drone just captured 800 photographs of a cornfield. Now what? This chapter walks you through the software that turns those photos into a single, usable map—and shows you how to do it without a computer science degree or a second mortgage.

You've flown the field. The microSD card has hundreds—maybe a thousand—individual photographs on it. Each photo covers a few acres from 250 feet, slightly overlapping its neighbors. Individually, they're useful for quick glances at specific spots. But the real power comes when they're stitched together into a single, seamless image of the entire field: an orthomosaic.

An orthomosaic is a geometrically corrected map made from overlapping aerial photos. The software identifies matching features across hundreds of images, calculates the precise position and orientation of each photo, corrects for lens distortion and camera angle, and blends them all into one continuous image that's geographically accurate—meaning every point in the image corresponds to a real GPS coordinate on the ground. You can measure distances, calculate areas, and overlay the map on other geographic data.

If you're flying with a multispectral camera, the software takes this a step further: it processes the separate wavelength bands, applies radiometric calibration, and generates vegetation index maps – NDVI, NDRE, and the other indices from Chapter 5 – as additional map layers.

The software landscape for drone mapping ranges from completely free to thousands of dollars a year. The good news is that for most agricultural applications, the free and low-cost options are genuinely capable. This chapter walks you through the options, helps you pick the right one for your situation, and takes you step-by-step through the processing workflow so you can go from a card full of photos to a finished map on your first try.

Understanding What the Software Produces

Before we compare software options, let's be clear about the products you're aiming for. Knowing what you need helps you avoid paying for capabilities you'll never use.

The Orthomosaic

This is the primary output and the one you'll use most. It's a single, high-resolution image of your entire field—essentially a giant aerial photograph, but geometrically corrected so that distances and areas are accurate. If you flew with an RGB camera, your orthomosaic is a natural-color image that looks like a satellite photo, but at dramatically higher resolution (inches per pixel instead of feet).

The orthomosaic is where you'll do most of your visual analysis: identifying stand gaps, weed patches, drainage patterns, equipment tracks, and visible stress. For many farmers, particularly those flying RGB, the orthomosaic is the only output they need.

Vegetation Index Maps

If you flew with a multispectral camera (or if your software generates enhanced-RGB indices), you'll get color-coded maps showing vegetation index values across the field. The most common are NDVI and NDRE, though your software may offer GNDVI, SAVI, and others from Chapter 5. These maps use a color gradient—typically red to green—to show low-to-high index values.

Index maps are where management decisions happen: identifying nitrogen-deficient zones for side-dress applications, detecting early stress for scouting prioritization, and tracking crop health changes across the season.

Elevation Models (Optional)

Most mapping software can also generate a Digital Surface Model (DSM) – a height map of the field showing elevation at every point. For agriculture, this is useful for drainage planning and identifying low spots. It's not essential for routine crop monitoring, but it's a nice bonus that comes free with most processing runs.

Prescription Files (Advanced)

Some agriculture-specific software can convert vegetation index maps into variable-rate prescription files – the .shp or .xml files that VRT equipment reads. This is the bridge between "seeing the problem" and "fixing the problem." Chapter 16 covers this process in detail. If prescription generation is important to you, it's a key differentiator when choosing software.

The Software Landscape: Your Options

We'll start with the free options, then work up to the paid platforms. For each, we'll cover what it does well, what it doesn't, and who it's best suited for.

Free and Low-Cost Options

DJI Fly App / DJI Terra (Lite)

Cost: Free with your DJI drone

What it does: The DJI Fly app includes a basic panorama and mapping function that can stitch images into a simple orthomosaic on your phone or tablet. DJI Terra's free tier offers more capable 2D mapping for DJI enterprise products like the Mavic 3 Multispectral.

Strengths: Zero cost, zero learning curve. You already have it installed. For a quick visual check – "let me see what the whole field looks like" – it's the fastest path from flight to image. No uploading to a cloud service, no waiting for processing queues.

Limitations: Limited processing power (phone/tablet hardware constrains quality), no vegetation index generation for consumer drones, no export in standard GIS formats, no prescription file creation. Fine for a quick look; not suitable for serious analysis.

Best for: Quick in-field visual checks. Your first few flights while you're learning. Situations where you need an answer in the field right now, not back at the office.

OpenDroneMap / WebODM

Cost: Free (open-source)

What it does: OpenDroneMap (ODM) is a fully capable, open-source photogrammetry engine that runs on your own computer. WebODM is a user-friendly browser-based interface for ODM that makes it accessible to non-programmers. Together, they can produce orthomosaics, elevation models, and vegetation index maps from both RGB and multispectral imagery.

Strengths: Completely free with no limitations on acreage, number of flights, or features. Your data never leaves your computer – no cloud uploads, no subscriptions, no corporate access to your field maps. Surprisingly capable for a free tool; the output quality is comparable to paid options for most agricultural applications. Active development community with regular updates.

Limitations: Steeper learning curve than commercial options. Requires a reasonably capable computer (16GB RAM minimum, 32GB recommended for large fields). Processing is slower than cloud-based services because it runs on your hardware. No mobile app – you process at home, not in the field. Limited agriculture-specific features like automatic prescription generation.

Best for: Farmers who are comfortable with computers and want a free, capable solution with no strings attached. Operations that value data privacy. Anyone processing RGB imagery who doesn't need prescription files.

Google Earth Pro

Cost: Free

What it does: Google Earth Pro isn't a mapping tool – it can't stitch your drone images together. But it's an excellent tool for viewing and comparing maps after they've been processed elsewhere. You can overlay your orthomosaic on Google's satellite imagery, compare your drone map to Google's lower-resolution view, and use its measurement and area tools to quantify what you're seeing.

Best for: A companion tool alongside your primary mapping software. Especially useful for comparing your field's appearance across multiple years using Google's historical imagery feature.

QGIS

Cost: Free (open-source)

What it does: QGIS is a full-featured Geographic Information System – professional-grade software for working with spatial data. It can display, analyze, and manipulate orthomosaics and index maps. You can overlay drone maps with soil maps, yield maps, SSURGO data, and other geographic layers. It can generate management zones and export prescription files.

Strengths: Extraordinarily powerful. Does everything that commercial GIS platforms costing thousands of dollars can do. Huge library of analysis tools and plugins. Can integrate drone data with every other type of spatial farm data.

Limitations: Significant learning curve. This is professional GIS software designed for professionals. The interface is not intuitive for newcomers. Like ODM, it doesn't stitch images – it works with maps that have already been created.

Best for: Farmers or agronomists who want to go deep on spatial analysis and are willing to invest time in learning. An excellent long-term skill that pays dividends well beyond drone mapping.

Paid Options Worth Considering

Pix4Dfields

Cost: $1,650/year (annual subscription) or $3,990 perpetual license

What it does: Pix4Dfields is designed specifically for agricultural drone data. It processes RGB and multispectral imagery into orthomosaics and index maps, generates management zones, and exports prescription files for VRT equipment. It runs locally on your computer – no internet required for processing.

Strengths: Agriculture-first design. The interface is built around the questions farmers actually ask: "where are my problem areas?" and "what prescription should I apply?" Excellent multispectral processing with proper radiometric calibration. Zone creation and prescription export in one integrated workflow. Fast local processing. Works offline after initial activation.

Limitations: $1,650/year is a meaningful investment. Requires a capable computer (similar to WebODM). Not cloud-based, so you cannot process from your phone. it's still $350 more than free. Requires a capable computer (similar to WebODM). Not cloud-based, so you can't process from your phone.

Best for: Farmers with multispectral cameras who need prescription files. The best value in the paid software market for agricultural applications.

DroneDeploy

Cost: $159/month (~$1,908/year) for Agriculture Lite (stand counts, plant health, Live Map); $299–$499/month ($3,588–$5,988/year) for standard plans

What it does: DroneDeploy is a cloud-based end-to-end platform: flight planning, image upload, processing, analysis, sharing, and annotation. It handles everything from mission planning through final map delivery. The processing happens on their servers, so your computer doesn't need to be powerful.

Strengths: The easiest platform to use, by far. If you can use a smartphone, you can use DroneDeploy. Cloud processing means no hardware requirements beyond a phone or tablet. Excellent mobile app for in-field viewing. Great sharing features – you can send a link to your agronomist, landlord, or insurance adjuster and they can view the map in a browser. Strong integration with many ag platforms.

Limitations: Standard plans are expensive – the annual cost exceeds the price of many drones. The Agriculture Lite tier ($159/month) makes cloud processing more accessible for individual farmers, though it has fewer features than standard plans. Requires internet connectivity for processing (not ideal in rural areas with poor service). Your data resides on their servers. Processing can be slow during peak season when many users are uploading simultaneously.

Best for: Service providers flying for multiple clients who value ease of use and sharing features. Large operations where the cost is a small fraction of the total precision agriculture budget. Farmers who want a "just works" experience and don't mind the premium.

Solvi

Cost: $600–$1,200/year

What it does: Solvi is a cloud-based platform focused on agriculture and environmental monitoring. It processes RGB and multispectral imagery, generates index maps, and offers annotation and comparison tools. Positioned as a middle ground between the simplicity of DroneDeploy and the specialization of Pix4Dfields.

Best for: Farmers who want cloud-based processing with agriculture-specific features at a lower price point than DroneDeploy. Good for operations that need sharing and collaboration features but can't justify DroneDeploy's premium pricing.

Sentera FieldAgent

Cost: Varies (often bundled with Sentera hardware)

What it does: Sentera's proprietary platform integrates tightly with Sentera multispectral sensors. It handles the full workflow from capture through prescription generation, with a focus on nitrogen management and crop health analytics.

Best for: Farmers already using Sentera hardware. The tight integration between sensor and software simplifies the multispectral workflow. Less useful if you're flying DJI or other non-Sentera platforms.

Software Comparison at a Glance

Software	Cost	RGB	Multispectral	Prescriptions	Runs On
DJI Fly / Terra	Free	Basic	Enterprise only	No	Phone/tablet
OpenDroneMap (WebODM)	Free	Full	Full	No (export to QGIS)	Your computer
Google Earth Pro	Free	View only	View only	No	Your computer
QGIS	Free	Full analysis	Full analysis	Yes (with plugins)	Your computer
Pix4Dfields	$1,650/yr	Full	Full	Yes	Your computer
DroneDeploy	$1,908–$5,988/yr	Full	Full	Yes	Cloud
Solvi	$600–$1,200/yr	Full	Full	Limited	Cloud
Sentera FieldAgent	Varies	Basic	Full (Sentera)	Yes	Cloud

OUR RECOMMENDATION

RGB-only farmers: Start with the DJI Fly app for in-field quick looks. Process full orthomosaics with WebODM when you need them.

Cost: $0.

Multispectral farmers (DIY): Pix4Dfields at $1,650/year. Best combination of agricultural features, processing quality, and value.

Service providers: DroneDeploy. The sharing and client management features justify the premium.

Power users: WebODM for processing + QGIS for analysis and prescriptions. Total cost: $0. Maximum capability.

The Processing Workflow: From Photos to Maps

Regardless of which software you use, the basic processing workflow is the same. Once you understand the steps, switching between platforms is straightforward – the buttons are in different places, but the logic is identical.

Step 1: Transfer Your Images

Remove the microSD card from your drone and insert it into your computer (or use the USB cable that came with the drone). Copy the images to a folder on your computer. For a 160-acre field at 250 feet, you'll have roughly 800–1,100 JPEG files totaling 5–15 GB depending on your camera's resolution.

Organization matters. Create a folder structure that you'll actually maintain. Something like:

SUGGESTED FOLDER STRUCTURE

Drone Flights /

2026 /

06-05_HomeField_V6_StandCount /

Raw Images /

Processed /

06-28_HomeField_V10_Nitrogen /

07-05_NorthFarm_R2_Disease /

Date, field name, growth stage, and purpose in the folder name.

You'll thank yourself in December when you're reviewing the season.

If you're using a cloud-based service like DroneDeploy, the upload happens directly from the app through your phone's cellular connection or over WiFi. For large flights, uploading over a rural cellular connection can be painfully slow. Many farmers find it faster to transfer files to a laptop in the field and upload from home WiFi in the evening.

Step 2: Create a New Project and Import

Open your mapping software and create a new project. In WebODM, this means clicking "New Project" and dragging your image folder into the browser window. In Pix4Dfields, you select "New Map" and point to the folder of images. In DroneDeploy, the app handles this during upload.

The software reads each image's embedded GPS data (latitude, longitude, altitude) and camera settings. It uses this metadata to establish the approximate position of each photo before the detailed alignment process begins.

Step 3: Process

This is the computationally intensive step where the software does its work. Click "Process" (or equivalent) and wait. What happens behind the scenes:

Feature detection. The software identifies distinctive features – corners, edges, texture patterns – in every image. In an agricultural field, these might include weed patches, soil marks, field edges, waterway boundaries, and even distinctive crop patterns.

Feature matching. The software finds features that appear in multiple overlapping images and uses them to calculate the precise relationship between images – how each photo connects to its neighbors.

Bundle adjustment. A mathematical optimization step that refines the position and orientation of every photo simultaneously. This is what makes the final map geometrically accurate.

Orthomosaic generation. The software projects each photo onto a common reference surface, corrects for lens distortion and camera angle, and blends overlapping areas into a seamless image.

Index generation (multispectral). If you flew with a multispectral camera, the software applies radiometric calibration (using your calibration panel images) and calculates vegetation indices for every pixel.

How Long Does Processing Take?

Platform	160 Acres (RGB)	160 Acres (Multispectral)	Notes
WebODM (16GB RAM laptop)	45–90 minutes	90–180 minutes	Depends heavily on computer specs
WebODM (32GB desktop)	20–45 minutes	40–90 minutes	Significant improvement with more RAM
Pix4Dfields (decent laptop)	20–40 minutes	30–60 minutes	Optimized for agricultural data
DroneDeploy (cloud)	30–60 minutes	45–90 minutes	Varies with server load; can be longer during peak season

Processing time scales roughly linearly with image count. A 40-acre field with 250 images processes about four times faster than a 160-acre field with 1,000 images. Plan accordingly—if you need a map by tomorrow morning, start processing after dinner, not after the late news.

Step 4: Review Your Orthomosaic

When processing completes, open the result and zoom in to check quality.

Look for:

Seamless stitching. Pan across the map at medium zoom. Do the crop rows line up continuously from one side to the other? Are there any visible seam lines where images were joined? Slight imperfections at the field edges are normal; gaps or warps in the middle indicate a processing issue.

Consistent exposure. Does the brightness look even across the field, or are there bright and dark patches that don't correspond to real crop differences? Exposure inconsistency – often caused by changing cloud cover during the flight – affects both visual quality and index accuracy.

Sharpness. Zoom in to full resolution in several places. Can you distinguish individual plants? See row spacing clearly? Blurry areas may indicate motion blur (wind), altitude variation, or a camera focus issue.

Complete coverage. Verify that the entire field is covered, including the edges and corners. Missing areas usually mean the flight boundary didn't extend far enough beyond the field edge.

Step 5: Generate Index Maps (Multispectral)

If you flew with a multispectral camera, most software generates index maps automatically as part of the processing step. If not, this is where you select which indices you want calculated. Based on Chapter 5:

Always generate: NDVI and NDRE. These cover 90% of agricultural management decisions.

Optionally generate: GNDVI (if you want additional chlorophyll data), SAVI (if this is an early-season flight with bare soil visible), or Chlorophyll Index (if your software supports it).

Set the color scale to a fixed range rather than auto-scaling, especially if you plan to compare maps across dates. For NDVI, a fixed range of 0.2–0.85 works well for Midwest row crops during the growing season. For NDRE, use 0.15–0.55.

Step 6: Export

Once you're satisfied with the map quality, export the files you need.

Common export formats:

GeoTIFF (.tif). The standard format for geospatial imagery. This is a regular image file with GPS coordinates embedded, so any GIS software can display it in the correct location. Export your orthomosaic and index maps as GeoTIFFs for maximum compatibility.

KML/KMZ. Google Earth format. Export as KMZ if you want to view your maps in Google Earth Pro. Convenient for quick visual comparison with Google's satellite imagery.

Shapefile (.shp). The standard format for management zone boundaries and prescription files. If you're creating VRT prescriptions, this is the format your applicator equipment expects.

PDF / JPG. If you just need a printable map for your records, scouting notes, or to share with a landlord, a high-resolution PDF or JPEG is the simplest option.

Step-by-Step: Your First Map With WebODM

Because WebODM is free, capable, and runs on your own computer, it's the best platform for learning. Here's a complete walkthrough for processing your first agricultural flight.

Installing WebODM

1. Go to https://www.opendronemap.org/webodm/ and download the installer for your operating system (Windows, Mac, or Linux).
2. Run the installer. WebODM uses Docker (a container system) to run the processing engine. The installer will prompt you to install Docker if you don't already have it. Follow the prompts – it's straightforward on Windows and Mac.
3. Launch WebODM. It opens in your web browser (typically at http://localhost:8000). You'll see a simple dashboard with a "New Project" button.

The installation process takes about 15 minutes on a good internet connection. You only do it once.

Processing Your First Flight

1. Click "New Project." Give it a name (e.g., "Home Field – V6 Stand Count – June 5") and click "Create."

2. Click "Select Images and GCP." Navigate to the folder containing your drone images and select all of them. For an 80-acre field, this will be around 400–500 files.

3. Review the default settings. For your first map, the defaults work fine. If you want to adjust, the key settings are: "Orthomosaic Resolution" (leave at default or set to 2–3 cm/pixel for agricultural use) and "Feature Quality" (set to "High" for best results, "Medium" if your computer is slower).

4. Click "Start Processing." WebODM begins working. You'll see a progress bar and status updates. Go do something else—processing a full field takes 30–90 minutes depending on your hardware.

5. View your results. When processing completes, the dashboard shows your orthomosaic on a map. Click on it to zoom in, pan around, and explore your field. The interface includes measurement tools (distance, area) and a 3D view option.

6. Download your outputs. Click "Download Assets" to get the GeoTIFF orthomosaic, elevation model, and any other outputs. These files are ready to open in Google Earth Pro, QGIS, or any other GIS software.

FIRST-MAP TIPS

Start with a small flight (40 acres or less) for your first processing run. This processes faster and lets you verify your workflow before committing to a large dataset.

If processing fails, the most common cause is insufficient RAM. Close other applications and try again, or reduce the "Feature Quality" setting.

Save your raw images permanently. Storage is cheap; re-flying a field at the same growth stage is impossible. Buy an external hard drive and back up every flight.

Computer Requirements: What Do You Need?

If you're processing locally (WebODM, Pix4Dfields, QGIS), your computer's specifications matter. Cloud-based services handle processing on their servers, so any modern phone, tablet, or laptop works for those.

Component	Minimum	Recommended	Notes
RAM	16 GB	32 GB	The single most important spec. More RAM = faster processing, larger datasets
Processor	Intel i5 / AMD Ryzen 5	Intel i7 / AMD Ryzen 7	Multi-core performance matters for photogrammetry
Storage	256 GB SSD	1 TB SSD	Raw images add up fast; SSD speed helps during processing
Graphics	Integrated OK	Dedicated GPU (4GB+)	Helpful for 3D rendering; not critical for basic processing
OS	Windows 10/11 or macOS	Windows 11 or macOS 13+	Linux also works (well) for ODM

If your current computer doesn't meet these specs, you have three options. First, use a cloud-based service – DroneDeploy, Solvi, or similar – where processing happens on their servers. Second, buy a refurbished business-class laptop with 32GB RAM; these run $400–$700 and are more than capable. Third, check if your local library, extension office, or community college has a computer lab with capable machines – some farmers have found this to be a workable solution for occasional processing.

Troubleshooting Common Processing Errors

Processing doesn't always work perfectly, especially in agricultural settings where uniform crop canopy makes feature matching more challenging. Here are the most common problems and their solutions.

Problem: Gaps or Holes in the Orthomosaic

Cause: Insufficient overlap, uniform canopy with too few distinctive features for the software to match, or wind-induced image blur at the field edges.

Fix: Increase overlap on your next flight (bump to 80%/75%). Ensure your flight boundary extends 100+ feet beyond the field edge. In extreme cases of uniform canopy (mid-season soybeans are the worst), try flying at a slightly lower altitude to capture more detail.

Problem: Wavy or Warped Map

Cause: The software didn't have enough altitude variation data to properly model the terrain, or GPS accuracy drifted during the flight. This is more common with drones that lack RTK correction.

Fix: In WebODM, try reprocessing with the "dsm" (Digital Surface Model) option enabled. In any software, ensure you're including the images from the edges of the field – sometimes excluding edge images causes distortion. If the problem persists, consider adding a few ground control points (visible markers on the ground with known GPS coordinates) for your next flight.

Problem: Exposure Banding (Bright/Dark Stripes)

Cause: Cloud shadows passing over the field during the flight created images with different exposure levels. The software blended them, but the brightness transitions are visible as stripes.

Fix: Avoid partly cloudy conditions for future flights (see Chapter 8). For the current dataset, some software allows you to manually remove the worst-affected images before reprocessing. In Pix4Dfields, the radiometric correction partially compensates for this. For RGB flights, there's no perfect fix after the fact.

Problem: Processing Fails Entirely

Cause: Usually a memory issue (not enough RAM), a corrupted image file, or too few usable images for the software to build a model.

Fix: Close all other applications and retry. Check that all image files open properly (corrupted files can crash the process). If using WebODM, reduce “Feature Quality” to “Medium” or “Low” to reduce memory demand. If the problem persists, try processing a subset of images (half the field) to determine whether it's a memory issue or a data quality issue.

Problem: Index Maps Look "Flat" or All One Color

Cause: The color scale is auto-ranging to a very narrow value range, making the entire field appear uniform. Or the radiometric calibration was incorrect (missing calibration panel images).

Fix: Set a fixed color scale (NDVI: 0.2–0.85; NDRE: 0.15–0.55) rather than auto-scaling. For multispectral data, verify that your calibration panel images were included in the processing batch. Some software requires you to identify the calibration images specifically.

Keeping It Simple: You Don't Need to Become a GIS Expert

After reading about GeoTIFFs, radiometric calibration, and bundle adjustment, you might feel like this is more complicated than you signed up for. Let us bring it back to earth.

For most farmers, the processing workflow is this: copy images from card to computer, open software, click "process," wait, look at the map. That's it. The software handles all the photogrammetry and math. You make management decisions based on what you see.

Here's what you actually need to be able to do:

Look at an orthomosaic and spot patterns. Streaks, patches, gradients, boundaries – the visual interpretation skills from Chapter 10. No software expertise required.

Compare two maps. Open this week's map next to last week's map and see what changed. Any software can do this.

Measure an area. How many acres are affected by that stress patch? Draw a boundary around it, and the software calculates the area. Useful for replant decisions, insurance claims, and spot-treatment planning.

Export a simple report. Generate a PDF or image of your map that you can print, email, or save for records.

If those four tasks are all you ever do with your maps, you're getting the full value of drone monitoring for a typical farm. The advanced capabilities – GIS analysis, prescription creation, multi-layer data integration – are there when you want them. But they're not prerequisites for getting started.

THE MINIMUM VIABLE WORKFLOW

Flight -> Copy images -> Process in WebODM -> Look at the map -> Walk the field

That's the workflow. Everything else is optimization. Start here. Add complexity only when you've identified a specific need that the basic workflow doesn't meet.

Chapter Summary

- An orthomosaic—a single, geometrically corrected map stitched from hundreds of overlapping photos—is the primary output from drone mapping software. For RGB farmers, it's often the only output you need.
- The software landscape ranges from completely free (WebODM, Google Earth Pro, QGIS) to $1,650/year (Pix4Dfields) to $1,908+/year (DroneDeploy Agriculture Lite) to $3,588+/year (DroneDeploy

standard plans). For most individual farmers, the free or low-cost options are genuinely sufficient.

- WebODM is the recommended starting point for local processing: free, capable, and your data never leaves your computer. Pix4Dfields ($1,650/year) is the best value for multispectral processing and prescription generation.
- The basic processing workflow is the same across all platforms: transfer images, create a project, process, review, and export.
- Processing a 160-acre field takes 20–90 minutes depending on software and hardware.
- Local processing requires a capable computer (16GB RAM minimum, 32GB recommended). Cloud-based services eliminate hardware requirements but add subscription costs and require internet connectivity.
- Common processing errors include gaps from insufficient overlap, exposure banding from cloud shadows, and flat index maps from incorrect calibration. Most issues are preventable through proper flight planning (Chapter 8).
- Organize your raw images in a clear folder structure (year/date_field_stage_purpose) and back up everything permanently. Storage is cheap; re-flying a field at the same growth stage is impossible.
- You don't need to become a GIS expert. The minimum viable workflow – process the images, look at the map, walk the field – delivers the full value of drone monitoring for most operations.

Notes and References

1. OpenDroneMap. "WebODM: Drone Mapping Made Easy." https://www.opendronemap.org/webodm/. Accessed February 2026.

2. Pix4D. "Pix4Dfields: Agriculture Drone Mapping Software." https://www.pix4d.com/product/pix4dfields. Accessed February 2026.

3. DroneDeploy. "Drone Mapping Software for Agriculture." https://www.dronedeploy.com/solutions/agriculture/. Accessed February 2026.

4. QGIS Development Team. "QGIS Geographic Information System." Open Source Geospatial Foundation. https://qgis.org/. Accessed February 2026.

5. Google. "Google Earth Pro." https://earth.google.com/. Accessed February 2026.

6. Solvi. "Drone Analytics for Agriculture." https://www.solvi.ag/. Accessed February 2026.

7. Sentera. "FieldAgent Platform." https://sentera.com/fieldagent/. Accessed February 2026.

8. Nex, F. and Remondino, F. "drone for 3D Mapping Applications: A Review." Applied Geomatics, 6(1), 1–15, 2014.

9. Turner, D., Lucieer, A., and Watson, C. "An Automated Technique for Generating Georectified Mosaics from Ultra-High Resolution Unmanned Aerial Vehicle (drone) Imagery." Remote Sensing, 4(5), 1392–1410, 2012.

10. Pix4D. "Image Quality for Photogrammetry: Requirements and Recommendations." Knowledge Base. https://support.pix4d.com/. Accessed February 2026.

Note: Software pricing, features, and system requirements referenced in this chapter are based on published information as of early 2026. The drone mapping software market evolves rapidly; verify current pricing and capabilities before purchasing. Free and open-source options listed (WebODM, QGIS, Google Earth Pro) have maintained their free status for many years and are expected to remain free.

Chapter 10: Reading Your Maps – Patterns, Problems, and Priorities

A drone map is not a diagnosis. It's a list of questions, ranked by urgency, with directions to the answers. The map tells you where to look. Your boots on the ground tell you what's wrong. This chapter teaches you to read the questions.

You've flown the field. You've processed the images. Now you're looking at a map—a beautifully detailed, high-resolution image of your entire field from above, possibly with color-coded index overlays showing relative crop health. The technology worked. But the technology is useless if you can't read what it's showing you.

Map interpretation is the most valuable skill in drone-based crop monitoring. More valuable than flying, more valuable than processing, more valuable than choosing the right camera. Because reading a map well—seeing the pattern, forming a hypothesis, and knowing where to go next—is what turns pixels into profit. A farmer who can read an RGB orthomosaic skillfully will get more value from a $1,200 drone than a farmer who can't interpret maps from a $6,000 multispectral system.

The good news is that you already have most of the skills you need. You know your fields. You know your soils. You know what your equipment does and doesn't do. You know the history of every acre you farm. What this chapter adds is a systematic framework for connecting what you see on the map to what you know about the field—and a disciplined process for confirming your interpretations before spending money on a response.

The Pattern Dictionary: What Common Shapes Mean

Crop stress and variability don't show up randomly on a map. They form patterns—and those patterns have causes. The shape, orientation, and distribution of what you see narrows the diagnosis before you ever leave the truck. Here's your field guide to the most common patterns and what they typically mean.

Streaks Parallel to the Planting Direction

What it looks like: Narrow strips of better or worse crop running in the same direction as your rows, typically repeating at regular intervals across the field.

Most likely cause: Planter issue. Every row unit on your planter has its own set of parts—seed disc, seed tube, closing wheels, downforce system. When one unit is worn or misadjusted, it produces a consistently different result from the others. The pattern repeats at the same interval as the problem: every 8th row on a 16-row planter means one unit is off. Every other row often points to a vacuum or pressure issue affecting alternating units.

What to check: Walk the affected rows at the beginning, middle, and end of the field. Compare plant spacing, emergence uniformity, and plant height to the adjacent healthy rows. Check your planter records – does the strip correspond to a specific row unit? Look for worn seed tubes, inconsistent downforce, or seed disc issues.

Financial impact: High. A single malfunctioning row unit on a 16-row planter affects 6.25% of every acre you plant with that machine. On 500 acres, that's 31 acres of compromised crop. At a conservative 20-bushel yield reduction on the affected rows, that's over $3,000 in corn alone.

Streaks Perpendicular to the Planting Direction

What it looks like: Bands of variation running across the rows, perpendicular (or at an angle) to the planting direction. The bands may alternate between better and worse crop, or you may see a single wide band that's different from the rest of the field.

Most likely cause: Applicator issue. Fertilizer spreaders, sprayers, and lime applicators travel across the field in a different direction from the planter. When a spreader has inconsistent coverage – overlap in some passes, gaps in others – the result is banding perpendicular to the rows. Sprayer nozzle issues (plugged, worn, wrong pattern) create similar perpendicular streaks.

How to narrow it down: Measure the width of the bands. If the bands match your spreader's effective swath width, that's your source. If they match your sprayer's boom width, check the sprayer. Compare the band timing to your application records – which product was applied in that direction?

Other possible causes: If the perpendicular band doesn't match equipment widths, consider hybrid or variety boundaries (if you planted different varieties in the same field), or previous-year harvest patterns (combine header width and direction).

Circular or Arc-Shaped Patterns

What it looks like: Curved lines, circles, or arcs of stressed crop, often near field corners or entry points.

Most likely cause: Compaction from equipment turns. Grain carts, semis, and combines make tight turns at field edges and entry points. Each turn compresses the soil in a circular arc. Repeated traffic on the same path—year after year—creates persistent compaction zones that restrict root growth and drainage. From the air, these arcs are unmistakable.

Other possible causes: Pivot irrigation patterns (if the field was previously irrigated or is near a pivot), or manure application turnaround areas.

What to check: Walk the arc with a penetrometer or simply push a tile probe into the soil. Compare the resistance in the arc to a spot 20 feet away. If compaction is the cause, the difference will be dramatic.

Linear Patterns Crossing the Field Diagonally or at an Angle

What it looks like: Straight lines that don't follow either the planting direction or the application direction. They may run diagonally, follow a gentle curve, or trace a path that has no obvious relationship to current farming operations.

Most likely causes: Tile lines (either functioning well and showing as better crop, or broken/failing and showing as stressed crop). Old fence rows or property boundaries that were removed decades ago but still affect soil structure. Buried utility lines. Natural soil transitions that follow glacial geology rather than field geometry.

Key diagnostic question: Is the linear pattern performing better or worse than the surrounding crop? A tile line typically shows up as a strip of better crop (better drainage = healthier roots) running through an area of poorer crop. An old fence row often shows up as a strip of different crop height—sometimes better (accumulated organic matter and nutrients along the fence line), sometimes worse (compaction from the old post holes and traffic lane).

What to do: Overlay the pattern on your tile map (if you have one), your county soil survey (available free through Web Soil Survey), or historical aerial photos (Google Earth's historical imagery often shows features that pre-date your ownership). If the line follows a buried tile, you've confirmed tile location – valuable information in itself. If it doesn't match anything, dig deeper. Literally, in some cases.

Random, Irregular Patches

What it looks like: Scattered areas of stressed or different-looking crop without an obvious geometric pattern. No repeating interval, no alignment with equipment passes, no clean linear boundaries.

Possible causes (ranked by frequency in the Midwest):

Localized compaction. Spots where heavy equipment sat, turned, or got stuck. Unlike the arcs from routine turns, these are one-off events – a grain cart that parked in the field during harvest, a truck that drove across the field to deliver tile, an excavator that worked on a waterway.

Disease pressure. Many foliar diseases spread in irregular patterns influenced by microclimate, spore dispersal, and canopy density. Tar spot in corn, frogeye leaf spot in soybeans, and Septoria in wheat all create patchy stress patterns that don't respect row geometry.

Weed escapes. Herbicide-resistant weed patches are irregular by nature – they establish where seeds land, which is influenced by wind, water flow, and combine chaff distribution. A waterhemp patch may follow a drainage path. A marestail patch may concentrate where combine chaff accumulated.

Wildlife damage. Deer feeding, wild turkey scratching, Canada geese grazing – these create irregular damaged areas, often concentrated near field edges, waterways, and woodlots.

Variable soil conditions. Gravel pockets, clay lenses, sand intrusions, old stream beds – natural soil variability creates patchy crop performance that looks random but is actually a reflection of the hidden geology beneath your field.

Edge Effects

What it looks like: A band of different crop performance running along one or more edges of the field, typically 30–200 feet wide.

Headland compaction. The most common edge effect. Every time you turn at the end of the field, the tractor, planter, sprayer, and combine drive over the same headland. Decades of this traffic creates a compacted zone 30–60 feet wide. The crop in the headlands often shows as slightly (or dramatically) stressed compared to the field interior.

Spray drift from neighbors. A band of curled, stunted, or discolored crop along a field boundary often indicates herbicide drift from an adjacent application. The damage typically fades with distance from the boundary. If it's on the downwind side, that's your confirmation.

Treeline shading. On the north or east side of a treeline, the crop receives less sunlight, especially early and late in the day. This creates a band of slightly lower vigor that widens through the season as the sun angle decreases.

Drainage concentration. Water from adjacent fields, roads, or developed areas may flow onto the edges of your field. The affected edge stays wetter longer, affecting early-season growth and root development.

Pattern Quick Reference

Pattern Shape	Orientation	Most Likely Cause	First Thing to Check
Regular narrow streaks	Parallel to rows	Planter row unit issue	Identify which row unit; walk the affected rows
Wide bands	Perpendicular to rows	Spreader/sprayer overlap or skip	Measure band width vs. equipment swath width
Alternating good/poor bands	Perpendicular to rows	Fertilizer spreader calibration	Match band pattern to spreader pass records
Arcs or circles	Near field edges/entries	Equipment turn compaction	Penetrometer test in arc vs. adjacent area
Straight diagonal lines (better crop)	Across field	Functioning tile lines	Overlay on tile map; confirm with probe
Straight diagonal lines (worse crop)	Across field	Broken tile, old fence row, utility	Check tile map; review historical aerials
Irregular scattered patches	Random	Disease, weeds, wildlife, compaction	Walk the patches; identify the specific cause

Irregular scattered patches	Random	Disease, weeds, wildlife, compaction	Walk the patches; identify the specific cause
Boundary band	Along field border	Headland compaction or spray drift	Compare to neighbor's application timing
Gradual gradient across field	Any direction	Soil type transition or drainage	Overlay on soil survey map
Small spots (< 0.5 acre)	Scattered	Ponding, old structures, localized issue	Historical aerials; walk the spots

Correlating With What You Know: Layering Information

A drone map in isolation is a picture. A drone map layered with your existing field knowledge becomes a diagnostic tool. The more context you bring to the interpretation, the faster you reach the right conclusion. Here's how to build that context.

Compare to Yield Maps

If you've been running a yield monitor for even a few seasons, you have a treasure trove of comparison data. Pull up this year's drone map alongside last year's yield map. Do the low-NDVI areas on the drone map correspond to the low-yield zones on the yield map? If yes, you're looking at a persistent problem—likely structural (compaction, drainage, soil type) rather than seasonal (weather, disease). Persistent problems are worth investing in because the fix pays off every year.

If the drone map shows stress in an area that typically yields well, you're looking at something new—a recent equipment issue, a new disease, a broken tile, a change in management. These new problems deserve immediate attention because the cause might be fixable this season.

Overlay on Soil Maps

Does the stress pattern follow soil type boundaries? In Western Ohio, where fields commonly contain three to five different soil series, soil transitions often explain crop variability. The USDA's Web Soil Survey (websoilsurvey.nrcs.usda.gov) provides free soil maps for every field in the country. Download your field's soil map and overlay it on your drone map.

If the stress pattern tracks the soil boundary—say, the low-vigor area matches the Pewamo silty clay loam and the healthy area matches the Crosby silt loam—you're seeing soil-driven variability. The management response might be drainage improvement, variable-rate seeding, or simply adjusting expectations for that zone.

If the stress pattern doesn't follow soil boundaries, soil type probably isn't the primary cause. Look elsewhere.

Check Your Application Records

As-applied maps from your planter, sprayer, and fertilizer applicator contain clues. If the stress pattern matches an application boundary – the edge of a prescription zone, the seam between two passes, the boundary of a different hybrid – the application itself is a suspect.

This is particularly relevant for hybrid or variety boundaries. If you planted two different corn hybrids in the same field and the drone shows a sharp transition in health at the boundary, the difference may be genetic, not agronomic. One hybrid may simply look different in an NDVI map without there being anything wrong with either. Check your planting records before assuming a problem.

Review Historical Aerial Imagery

Google Earth Pro's historical imagery slider is one of the most underused tools in agriculture. You can view satellite images of your field going back to the early 2000s in many areas – sometimes earlier. Patterns that appear in historical imagery are almost certainly structural (soil, drainage, buried features). Patterns that appear for the first time in this year's drone imagery are likely new problems.

Historical imagery is especially powerful for identifying old fence rows, buried structures, former building sites, and drainage features that pre-date your memory of the field. We've seen farmers identify features visible in 2005 imagery that explained yield variability they'd been puzzling over for a decade.

THE FIVE-LAYER APPROACH

For any stress pattern you identify on a drone map, layer these five sources of information:

1. The drone map itself (shape, location, severity of the pattern)

2. Yield map history (is this persistent or new?)

3. Soil survey (does it follow soil type boundaries?)

4. Application records (does it match an equipment pass or prescription boundary?)

5. Historical imagery (has this pattern appeared before?)

Most patterns can be diagnosed – or at least narrowed to two or three likely causes – using these five layers without ever leaving your desk. The ground-truthing step confirms which cause is correct.

False Alarms and Limitations: When the Map Lies

Not every color difference on your map represents a real crop problem. Learning to distinguish genuine stress from artifacts and noise is essential for avoiding wasted time and unnecessary expense. Here are the most common false alarms.

Cloud Shadows

We've mentioned this in previous chapters, but it deserves emphasis here because cloud shadows are the single most common false alarm in drone maps. A cloud shadow creates a dark patch that looks indistinguishable from crop stress in an RGB orthomosaic. In an NDVI map, the reduced light in the shadow area produces artificially low index values.

How to identify: Cloud shadow patterns usually have soft, irregular edges that don't follow field features. They may cut across rows, soil boundaries, and management zones with equal indifference. If you flew on a partly cloudy day and see patches that don't correspond to anything you know about the field, suspect clouds first. Compare the pattern to known field features—if it doesn't align with anything, it's probably a shadow.

Temporary Moisture Patterns

If you flew within 24–48 hours after a significant rain, you may see patterns that reflect where water is standing or where the soil surface is still saturated. These patterns will disappear as the field dries. They're not meaningless—they tell you where water goes, which is valuable information—but they don't necessarily indicate crop health issues. A field that looks "stressed" 12 hours after a 2-inch rain might be perfectly healthy two days later.

When it matters: If the same moisture pattern shows up flight after flight, the drainage issue is real and persistent. That's different from a one-time snapshot of temporary ponding.

Crop Variety Differences

If you planted different hybrids or varieties in the same field, they may have different NDVI signatures at the same growth stage—even if both are perfectly healthy. One corn hybrid may have a slightly more upright leaf architecture that reflects NIR differently. One soybean variety may mature three days earlier, changing its spectral appearance in a late-season map.

How to identify: If the "stress" boundary is a perfectly straight line that coincides with your hybrid changeover point, it's almost certainly a variety difference, not a management problem. Check your planting records.

Stitching Artifacts

The software that creates your orthomosaic sometimes struggles where images join, particularly over uniform crop canopy. You may see slight color shifts, blurred areas, or "ghost" patterns at image boundaries. These are processing artifacts, not crop problems.

How to identify: Stitching artifacts tend to follow straight lines at odd angles that don't correspond to any field feature. They may appear as narrow bands of slightly different color or sharpness running across the map. If a pattern has perfectly straight edges at a 30-degree angle to everything in the field, it's probably a seam line in the orthomosaic.

Dew and Time-of-Day Effects

Early morning flights when dew is still on the canopy can produce dramatically different reflectance values than midday flights of the same field. The water droplets on the leaves alter both visible and NIR reflectance. If you're comparing two maps of the same field and one looks markedly different, check the flight times. A 7:00 AM flight and a noon flight aren't directly comparable.

THE GOLDEN RULE OF MAP INTERPRETATION

Don't chase every yellow spot. Look for patterns that are large enough to matter economically, persistent across multiple flights, and consistent with something you know about the field. A 0.3-acre yellow patch in the corner of a 160-acre field is almost never worth investigating. A 15-acre band of reduced NDRE running through the middle of the field is always worth investigating.

Ground-Truthing: Where Maps Meet Mud

Ground-truthing is the non-negotiable bridge between map observation and management action. No matter how clear the pattern, no matter how confident you are in your diagnosis, you verify it on the ground before spending money on a response. Every time.

This discipline saves you from two expensive mistakes: treating the wrong problem (the yellow band was a variety boundary, not nitrogen deficiency) and treating a problem that doesn't exist (the dark patch was a cloud shadow, not disease).

The Ground-Truthing Process

7. Mark your targets on the map. Before you leave the house, identify the specific spots you need to visit. Most mapping software lets you drop pins or waypoints. Mark the center of the problem area and at least one comparison point in a healthy area nearby. You want to see the contrast between the problem and the baseline.

8. Navigate to the spot. If your software exports waypoints to your phone's GPS, use that to navigate. Otherwise, use landmarks visible in both the drone map and the real world—tree lines, waterways, tile risers, field edges—to orient yourself. It's surprisingly easy to get turned around in the middle of a 160-acre cornfield at VT.

9. Observe before diagnosing. When you reach the target area, slow down. Don't immediately start looking for the cause you expect to find. Instead, observe the general condition: plant height, color, canopy density, uniformity. How does it compare to the healthy comparison area?

Is the difference as dramatic as the map suggested, or more subtle?

10. Examine individual plants. Pull a plant from the stressed area and one from the healthy area. Compare roots (depth, density, color), stalks (diameter, firmness), and leaves (color, lesions, curling, insect damage). Smell the roots – certain diseases have distinctive odors. Split the stalk if you suspect disease.

11. Check the soil. Push a tile probe or soil probe into the ground. Is there a compaction layer? Is the soil wet or dry compared to the healthy area? Dig a shovel slice and look at root penetration depth.

12. Document everything. Photograph the plants, the soil, and the general area. Drop a GPS pin from your phone. Write down what you found. This documentation serves three purposes: it supports your management decision, it builds your field history, and it provides evidence for insurance or warranty claims if relevant.

The Ground-Truthing Checklist

GROUND-TRUTHING FIELD FORM (LAMINATED VERSION IN APPENDIX E)

- [ ] Date: ______ Field: ______
  Location (GPS or
  description): ______
- [ ] Map observation: What did the map show? (color, pattern, index value)
- [ ] Visual assessment from road/edge: Does the problem look visible from the field edge?
- [ ] In-field observation: Plant height, color, canopy density vs. comparison area
- [ ] Plant examination: Roots, stalk, leaves – any abnormalities?
- [ ] Soil examination: Moisture, compaction, root depth, odor
- [ ] Comparison to healthy area: What's different?
- [ ] Preliminary diagnosis: What do you think caused the pattern?
- [ ] Action needed? (Yes/No/Monitor) If yes, what and when?
- [ ] Photos taken? GPS pin dropped?

Prioritizing What to Investigate: The Triage System

A drone map of a full-season cornfield will almost always show some variability. No field is perfectly uniform. The skill isn't finding problems—it's deciding which problems are worth your time and money.

Use this triage system to prioritize your ground-truthing and response:

Priority	Criteria	Action	Example
Immediate	Severe stress (NDRE < 0.20), large area (>10 acres), time-sensitive window	Ground-truth today; act this week if confirmed	Nitrogen deficiency at V10 with side-dress window closing
Soon	Moderate stress (NDRE 0.20–0.35), medium area (3–10 acres), actionable	Ground-truth within 3–5 days; plan response	Disease onset in soybeans at R3 with fungicide window open
Monitor	Mild variability (NDRE 0.35–0.40), small area (<3 acres), limited action	Note it; re-fly next scheduled flight; compare	Slightly lower vigor in corner; cause unclear
Record	Any pattern of interest, not currently actionable	Document for end-of-season review and future planning	Compaction patterns for fall tillage planning

The key factors in prioritization are size (how many acres are affected), severity (how far below normal are the index values), timing (is there still a management window open), and actionability (do you have the means to respond). A severe problem on 20 acres with a treatment window still open is always your first priority. A mild issue on 2 acres with no available response goes in the "record" bin.

Case Study: From Map to Diagnosis to Solution

To bring all of these concepts together, let's walk through a real-world case study—a problem identified on a drone map, diagnosed through ground-truthing, and resolved through targeted management action.

The Map

A Darke County farmer flew his 140-acre cornfield at V10 using a DJI Mavic 3 Multispectral. The NDRE map showed healthy crop (0.42–0.50) across most of the field, with one exception: a wedge-shaped zone in the northeast quadrant, roughly 18 acres, showing NDRE values of 0.25–0.33. The zone had a sharp boundary on its western edge and a gradual fade on its eastern edge.

The Desk Analysis

Before heading to the field, the farmer checked his records. The wedge shape didn't match any equipment pass—it wasn't perpendicular to rows (ruling out spreader issues) and wasn't parallel to rows (ruling out planter issues). He overlaid the pattern on his yield map from the previous year. The same area had yielded 15–20 bushels below the field average in each of the past three years, but he'd attributed that to soil type—the area was on a Miamian silt loam while the rest of the field was Crosby.

However, the NDRE values were lower than he'd expect for a soil-type difference alone. Soil type might reduce vigor by 5–10%, but the NDRE drop was 30–40% below the field average. Something else was going on.

The Ground-Truthing

He drove to the northeast corner and walked into the wedge zone. The corn was visibly shorter – perhaps 18 inches less than the healthy crop 200 feet west. The lower leaves were yellowing. Classic nitrogen deficiency symptoms. But the field had received a uniform 180 lbs of nitrogen as anhydrous ammonia.

He pulled several plants and examined the roots. In the stressed zone, roots were shallow – barely penetrating beyond 6 inches. In the healthy zone, roots were past 12 inches. He pushed a tile probe in both areas. The stressed zone had a hard layer at about 8 inches; the healthy zone had no resistance to 18 inches.

The diagnosis crystalized: compaction. The northeast quadrant was where the grain cart had entered the field for the past several years, making repeated turns as it shuttled between the combine and the road. The compaction restricted root growth, which restricted nitrogen uptake, which produced the nitrogen-deficiency symptoms and the low NDRE values. The nitrogen was in the soil—the roots couldn't reach it.

The Response

The problem wasn't fixable mid-season—you can't correct compaction in standing corn. But the diagnosis changed three future decisions:

1. That fall, he deep-ripped the 18-acre compacted zone after harvest. Cost: approximately $400 for custom deep tillage.
2. He rerouted his grain cart traffic to a designated lane on the opposite side of the field, using a gravel access point he'd been meaning to install. Cost: approximately $2,500 for the gravel pad and field entrance.
3. He began using his drone map as a baseline to monitor the zone's recovery in subsequent years, flying the same field at V10 each season.

The Result

The following year's V10 NDRE map showed the wedge zone at 0.38–0.44 – not fully recovered, but markedly improved. Yield in the zone increased by an estimated 22 bushels per acre. On 18 acres of corn at $4.50 per bushel, that's roughly $1,780 in recovered yield – in the first year. The deep tillage and gravel pad investments were recovered within two seasons.

Without the drone map, the farmer would have continued attributing the underperformance to soil type—a fixed condition he assumed was beyond his control. The map showed him the pattern was too severe for soil type alone, the ground-truthing revealed compaction as the actual cause, and the targeted response turned an 18-acre drag on his field into a productive zone.

LESSONS FROM THE CASE STUDY

The drone map didn't diagnose the problem. It identified the pattern and quantified the severity. The ground-truthing diagnosed the problem. The combination of aerial view and boots-on-the-ground examination is what produced the actionable answer.

Historical context (yield maps) helped distinguish between "always been that way" and "worse than it should be." Both are true – the area always underperformed, but the degree of underperformance pointed to a fixable cause.

The financial return ($1,780+/year in recovered yield) justified the combined investment (drone flight time + deep tillage + gravel pad) within two seasons. The drone didn't create the value – it revealed where the value was hiding.

Map Interpretation Worksheet

Use this worksheet each time you review a drone map. A printable version is included in Appendix F.

MAP INTERPRETATION WORKSHEET

Date: ______ *Field:* ______ *Flight*

> type: RGB / Multispectral

> Growth stage: ______ Index viewed:

______ Scale range:

______

Step 1: Overall Assessment

What percentage of the field looks healthy? ______%

How many distinct problem areas do you see? ______

Step 2: Pattern Analysis (for each problem area)

Location in field: ______ Approximate size (acres):

______

Shape: Streaks / Bands / Patches / Arcs / Gradient / Edge effect

> Orientation: Parallel to rows / Perpendicular / Diagonal / Random

Severity (index value or visual): ______

Step 3: Correlation Check

Matches yield map history? Yes / No / Partially

Follows soil type boundary? Yes / No

Matches equipment pass? Yes / No / Which equipment?

______

Visible in historical imagery? Yes / No / First occurrence

Step 4: Hypothesis and Priority

Most likely cause: ______

> Priority: Immediate / Soon / Monitor / Record

>

> Ground-truthing needed? Yes / No Date planned:

______

Chapter Summary

- Map interpretation is the highest-value skill in drone-based monitoring. The shape, orientation, and distribution of patterns narrow the diagnosis before you leave the truck.
- Six common pattern types: parallel streaks (planter), perpendicular bands (applicator), arcs (compaction turns), diagonal lines (tile/fencerows/soil), random patches (disease/weeds/wildlife), and edge effects (headlands/drift/shading).
- The five-layer approach combines the drone map with yield map history, soil survey data, application records, and historical aerial imagery to diagnose problems from your desk.

- Common false alarms include cloud shadows, temporary moisture patterns, crop variety boundaries, stitching artifacts, and dew/time-of-day effects. Learn to recognize these before chasing them to the field.
- Ground-truthing is non-negotiable. Maps tell you where to look; boots on the ground tell you what's wrong. Walk the spots, examine plants and soil, document findings, and confirm the diagnosis before spending money.
- Use the triage system to prioritize: immediate (severe, large, time-sensitive), soon (moderate, medium, actionable), monitor (mild, small, unclear), record (note for future planning).
- Layer context to strengthen interpretation: persistent patterns visible in yield maps point to structural causes; new patterns visible only this season point to recent problems.
- The case study demonstrates the full workflow: map identification, desk analysis, ground-truthing, diagnosis (compaction masquerading as nitrogen deficiency), targeted response (deep tillage + traffic rerouting), and measurable financial return ($1,780/year on 18 acres).

Chapter 10 Notes and References

1. Shanahan, J.F., Schepers, J.S., Francis, D.D., et al. "Use of Remote-Sensing Imagery to Estimate Corn Grain Yield." Agronomy Journal, 93(3), 583–589, 2001.

2. Mulla, D.J. "Twenty-Five Years of Remote Sensing in Precision Agriculture." Journal of Environmental Quality, 42(2), 358–377, 2013.

3. USDA Natural Resources Conservation Service. "Web Soil Survey." https://websoilsurvey.nrcs.usda.gov/. Accessed February 2026.

4. Google. "Google Earth Pro: Historical Imagery." https://earth.google.com/. Accessed February 2026.

5. Blackmer, T.M., Schepers, J.S., Varvel, G.E., and Walter-Shea, E.A. "Nitrogen Deficiency Detection Using Reflected Shortwave Radiation from Irrigated Corn Canopies." Agronomy Journal, 88(1), 1–5, 1996.

6. Hamza, M.A. and Anderson, W.K. "Soil Compaction in Cropping Systems: A Review of the Nature, Causes, and Possible Solutions." Soil and Tillage Research, 82(2),

121–145, 2005.

7. Ohio State University Extension. "Soil Compaction: Causes, Effects, and Management." https://extension.osu.edu/. Accessed February 2026.

8. Hatfield, J.L. and Prueger, J.H. "Value of Using Different Vegetative Indices to Quantify Agricultural Crop Characteristics." Remote Sensing, 2(2), 562–578, 2010.

9. Thomason, W.E., Phillips, S.B., Davis, P.H., et al. "Early Season Nitrogen Monitoring with Active Sensors in Winter Wheat." Precision Agriculture, 12, 756–768, 2011.

10. Kyveryga, P.M., Blackmer, T.M., and Mueller, D.S. "Interpreting Aerial Photographs for Crop Scouting." Iowa State University Extension, PM 1871, 2009.

Note: Pattern interpretations described in this chapter represent the

most common causes in Midwest row crop agriculture. Individual field conditions, cropping systems, and management histories may produce patterns with different causes. Ground-truthing is always required to confirm diagnoses. The case study is based on a real farm situation with details modified for privacy.

Chapter 11: Your First Season—A 12-Week Action Plan

You've read the book. You understand the science, the operations, and the applications. Now it's time to do it. This chapter condenses everything into a concrete, week-by-week plan that takes you from unboxing to first-season results in twelve weeks.

Twelve weeks is one growing season from pre-plant through mid-grain-fill. It's also all the time you need to go from zero drone experience to a farmer who flies, processes, interprets, and acts on aerial data. This timeline assumes you're starting with an RGB drone (Chapter 2's $1,200–$1,500 recommendation) and building toward the monitoring program described throughout this book.

The plan is structured in four phases: Pre-Season Preparation (Weeks 1–4), Early Season Flights (Weeks 5–8), Mid-Season Critical Decisions (Weeks 9–12), and a Late-Season Wrap-Up that carries you through harvest and into winter planning. Each week has specific tasks, a time estimate, and clear outcomes.

If you're reading this mid-season and can't start from Week 1, jump to whatever phase matches your current growth stage. Any entry point is better than waiting until next year.

Phase 1: Pre-Season Preparation (Weeks 1–4)

This phase happens 4–6 weeks before planting. You're ordering equipment, setting up, and practicing before the crop goes in the ground. The goal is to be comfortable and confident with the drone before you have real crop data to worry about.

Week 1: Decide and Order

Task Time Details

Choose your drone 1–2 hours Refer to Chapter 2. For first season: DJI Mini 4 Pro ($1,200–$1,500). Includes camera, controller, and three batteries.

Order the drone and 30 Extra SD cards (128 GB min recommended), spare accessories propellers, carrying case if not included, landing pad ($15–20)

Create FAA account 30 min Visit faadronezone.faa.gov. Registration and register drone costs $5 for 3 years. Print your registration number.

Begin TRUST test one hour Free online test at trust.modelaircraft.org (if not Part 107) or equivalent. Required for recreational flyers. Takes 30–60 min.

Download mapping 15 min DroneDeploy (free tier) or Pix4Dcapture on app your phone/tablet. Create account.

WEEK ONE INVESTMENT

Drone + accessories: $1,250–$1,600

FAA registration: $5

TRUST test: Free

Time: 3–4 hours

Week 2: Unbox, Set Up, and Learn the Controls

Task Time Details

Unbox and charge one hour Fully charge drone and controller. Update all batteries firmware through the DJI app (requires Wi-Fi).

Attach registration five Label maker or min permanent marker on a small

number to drone sticker. Must be visible.

Read the 30 min Don't skip this. Understand return-to-home,

quick-start guide obstacle avoidance, and emergency

procedures.

First flight: hover 30 Calm day. Open area. min Just take off, hover at

test in your yard 10 feet, practice gentle movement, land. Get

or driveway the feel.

Second practice: 45 min Fly a rectangle at 50 feet. Practice smooth

basic navigation turns, altitude changes, and the camera

gimbal. Fly away from yourself and back.

Set up folder 20 min Create the folder structure from Chapter 18:

structure on Farm_Drone_Data / 2026 / [Field names]

computer

THE MOST COMMON WEEK 2 MISTAKE

Flying too far, too fast, too high on your very first flight. Start in your driveway or a mowed area. Hover at 10 feet. Get comfortable with the controls responding to your inputs before you send the drone 400 feet over a neighbor's field. The first hour of flying should be boring. Boring means safe.

Weeks 3–4: Practice Flights Over Empty Fields

Task Time Details

Practice flight 1 hour Fly at 200–250 feet over a field you know

over an empty field well. Practice smooth, overlapping flight

(manual) lines. Take photos manually.

First automated 1.5 hours Use DroneDeploy or Pix4Dcapture to plan and

mapping mission execute an automated survey of 20–40 acres.

Let the software fly the drone.

Process your first two hours Upload images to DroneDeploy (cloud) or

orthomosaic Pix4Dfields (desktop). Follow the processing

steps from Chapter 9.

Review the map 30 min Look for stitching errors, gaps, blur.

critically Compare the map to what you know the field

looks like. Is it accurate?

Repeat: fly a two hours Repetition builds confidence. By the end of

second field, Week 4, the workflow should feel routine.

process, review

Check your B-VLOS 30 min Review Chapter 8's pre-flight checklist.

and weather Practice evaluating wind, clouds, and

assessment skills airspace before every flight.

By the end of Week 4, you should have 3–5 hours of flight time, have processed 2–3 orthomosaics, and feel comfortable with the flight-to-map workflow. You've done the hard part—the learning curve is behind you. Everything from here is applying that skill to real crops.

End of Phase 1 Checkpoint

- [ ] Drone registered with FAA and labeled
- [ ] TRUST test passed (or Part 107 obtained)
- [ ] 3–5 hours of flight practice completed
- [ ] 2–3 practice orthomosaics processed successfully
- [ ] Mapping app comfortable on automated missions
- [ ] Folder structure created on computer

- [ ] Pre-flight checklist memorized

Phase 2: Early Season Flights (Weeks 5–8)

The crop is in the ground and growing. Your first real agricultural flights deliver immediate, tangible value—stand assessment, emergence evaluation, and early weed scouting. These are all RGB applications; no multispectral camera needed.

Weeks 5–6: First Crop Flight – Stand Assessment (V2–V4)

Task Time Details

Fly your 1–1.5 Automated mission at 200 ft AGL. This is

highest-value field your first real crop hours map.

at V2–V4

Process the 1–2 hours Same workflow you practiced in Weeks 3–4.

orthomosaic Processing time varies by field size.

Evaluate emergence 30 min Look for: skips, doubles, uneven emergence,

and population planter row gaps. Refer to Chapter 13.

Identify replant 30 min Storms, frost, or crusting may have created

candidates (if areas needing evaluation. Use Chapter 13's

applicable) replant decision framework.

Ground-truth one or one Walk the areas that hour looked different on the

two areas map. Count plants. Confirm what you see.

Record observations 15 Create notes.txt in the min flight folder. What

did you see? What decisions did the map

inform?

This first crop flight is often the moment when the drone's value becomes real. You'll see things you didn't know were there—a planter skip you missed, an emergence gap from a cold pocket, or a tile blowout that's been there for years but you never noticed from the road. That first "I didn't know that" moment is what converts skeptics into advocates.

Week 7: Review, Ground-Truth, and Document

Task Time Details

Review your first 30 min Look at it again with fresh eyes. What

crop map in detail patterns do you see? Any surprises?

Overlay with soil 30 min Download soil survey from Web Soil Survey.

map (optional) Compare stress patterns to soil type

boundaries.

Walk additional 1–2 hours Focus on areas where the map shows something

areas identified in unexpected. Bring the map on your phone for

the map reference.

Share the map with 30 min Getting a second set of eyes on your first

your agronomist or map accelerates your interpretation

neighbor learning.

Make one management Even a small one varies counts: adjusting planter

decision based on settings, flagging an area for herbicide,

the map scheduling a tile check.

Week 8: Second Flight – Baseline Health (V6–V8)

Task Time Details

Fly the same 1–1.5 Same flight plan, same altitude.

field(s) again at hours Consistency enables comparison.

V6–V8

Process the 1–2 hours Standard workflow.

orthomosaic

Compare to the 30 min Did the emergence gaps persist? Did the

V2–V4 map thin areas fill in? Any new problems?

Begin early weed 30 min Before canopy closure, weed patches are

scouting from the visible. Mark for spot treatment.

map

Fly 1–2 additional 2–3 hours Expand to more fields now that the workflow

fields is comfortable.

End of Phase 2 Checkpoint

- [ ] At least 2 crop flights completed on your primary field
- [ ] Stand assessment completed; planter performance evaluated
- [ ] At least one ground-truthing walk completed
- [ ] At least one management decision informed by drone data
- [ ] Flight-to-map comparison (V4 vs. V8) reviewed
- [ ] Expanded to at least 2–3 fields

Phase 3: Mid-Season Critical Decisions (Weeks 9–12)

This is when drone data delivers the highest return. The crop is establishing its yield potential, and decisions made now—side-dress nitrogen, fungicide timing, scouting priorities—have the largest financial impact. If you have multispectral capability, this is the phase where NDRE becomes essential. If you're flying RGB only, your maps still guide targeted scouting and visual health assessment.

Weeks 9–10: The Critical NDRE Flight (V8–V10)

Task Time Details

Fly all corn fields 2–4 hours V8–V10 is the nitrogen assessment sweet

(NDRE if available, spot. Prioritize fields with side-dress

RGB if not) planned.

Process NDRE maps 2–4 hours If multispectral: generate NDRE index map.

(or RGB If RGB: look for color differences, canopy

orthomosaics) gaps.

Identify areas of 30 min Low-NDRE zones (if multispectral) or

concern on the map pale/thin canopy areas (if RGB).

Ground-truth the 1–2 hours Walk them. Confirm nitrogen deficiency vs.

low-vigor areas other causes (Chapter 12's diagnostic

flowchart).

If applicable: 1–2 hours Chapter 16's workflow. Even a simple 3-zone

create zones for map improves on a flat rate.

VRT side-dress

Share maps with 30 min Email or print the map with your

agronomist for observations. Get their interpretation

input before deciding.

IF YOU'RE RGB-ONLY: YOU'RE STILL IN THE GAME

Don't skip mid-season flights because you don't have a multispectral

camera. RGB imagery at V8–V10 reveals color variation

(nitrogen-stressed corn is lighter green), canopy density

differences, weed escapes, and equipment damage. You won't generate

NDRE-based prescriptions, but you'll identify where to scout and make

a better flat-rate decision than you would without the map. That

alone is worth the flight.

Week 11: Act on the Data

Task Time Details

Make nitrogen varies Using Weeks 9–10 maps + ground-truthing.

side-dress decision Flat rate or VRT prescription.

(if applicable)

Execute spot varies Spot-spray weed escapes identified from map.

applications (if Targeted scouting of disease suspects.

applicable)

Update scouting 30 min Focus your walking time on the areas the map

priorities based on flagged. Skip the parts that look healthy.

map

Fly soybean fields 1–2 hours Soybeans now entering the period when

(R1–R2) disease, SCN, and SDS become visible.

Process and review 1–2 hours Look for the patterns from Chapter 15: SDS

soybean maps patches, uneven canopy, iron chlorosis.

Week 12: Third Major Flight – Health Assessment (VT–R2 Corn / R3 Soybeans)

Task Time Details

Fly all fields at 3–5 hours This is the last high-value flight before

VT–R2 (corn) / grain fill. Disease pressure peaks now.

R2–R3 (soybeans)

Process all maps 3–5 hours Batch process. Prioritize fields where

you're considering fungicide.

Compare to Weeks 30 Is stress expanding, stable, or recovering?

9–10 maps min/field New patterns = scout immediately.

Make fungicide varies Three conditions: disease-consistent

decision using pattern + confirmed disease + within

Chapter 15 application window.

framework

Document everything 30 This flight's maps are min the data you'll compare to yield at harvest. Notes matter.

End of Phase 3 Checkpoint

- [ ] Critical V8–V10 nitrogen flight completed (corn)
- [ ] Ground-truthing confirmed (or ruled out) nitrogen issues
- [ ] At least one input decision (N rate, fungicide, spot spray) informed by drone data
- [ ] VT–R2 health assessment completed on all key fields
- [ ] Soybean fields flown and evaluated
- [ ] Maps compared between flights to track change over time

Phase 4: Late Season and First-Year Wrap-Up

After Week 12, the crop is filling grain and your drone monitoring program shifts from active decision-making to documentation, evaluation, and planning. This phase extends through harvest and into winter.

Late Season: R3 Through Pre-Harvest

Task Time Details

Optional flight at 1–2 hours Late-season stress assessment. Documents R4–R5 (corn) stalk integrity, late disease, harvest timing zones.

Optional flight at 1–2 hours Maturity assessment. Identifies R5–R6 (soybeans) early-maturing zones for harvest sequence planning.

Fly any fields with Insurance varies documentation: fly within 48 hours suspected if possible. Save raw images and hail/storm damage orthomosaic.

Pre-harvest 1–2 hours Last chance to document drainage patterns

drainage assessment before harvest. Wet areas guide winter tile

(if recent rain) planning.

Harvest: Collecting the Answer Key

Harvest is when you get the answer key. Your yield monitor records what actually happened—the outcome that all your drone monitoring was trying to predict and improve. Saving yield data is not optional in this program.

Save yield data from every field. Export from the combine monitor and store in your drone data folder structure alongside the aerial maps. Name consistently: 2026_North80_yield_map.

Note anything unusual during harvest. Lodging areas, disease patches, wet spots that slowed the combine, areas that yielded surprisingly well or poorly. These observations ground-truth your season-long drone monitoring in reverse.

Winter: The Most Important Work of the Season

This is when you close the loop. The winter review transforms a first season of data collection into a permanent improvement in your management. Invest one afternoon in this review; it will shape next year's entire program.

Winter Review Time What You're Learning

Task

Overlay drone maps 1–2 hours Did mid-season stress (low NDRE or RGB

with yield maps color) predict low yield?

(Chapter 18)

Identify persistent 30 Persistent problems need structural fixes

vs. one-time min/field (drainage, variety, soil). One-time

problems problems may not recur.

Calculate one hour Total cost (equipment + time) vs. total
first-year ROI value of decisions drone data informed. Be
honest.

Document what 30 min Which flights were most useful? Which were
worked and what a waste of time? What would you do
didn't differently?

Plan next year's 30 min Block the key growth stages on your
flight schedule calendar now. What fields, what stages,
what questions?

Decide on equipment 30 Is multispectral min justified? Do you need
upgrades (if any) more batteries? A better computer? Or is
current setup fine?

Your First Crop Flight Day: Detailed Schedule

Here's what a successful first crop flight day looks like, step by step. Print this page and bring it to the field.

Time	Activity	Notes
Before leaving the house	Check weather: wind < 15 mph, no rain, no storms in forecast	Use Windy app. Postpone if conditions are marginal.
Before leaving the house	Charge all batteries (drone + controller + phone/tablet)	Start the night before. Full charge on everything.
Before leaving the house	Clear SD card and verify storage space	Format the card in the drone, not on the computer.
Arrive at field	Park on flat, dry, stable surface away from power lines and trees	Use a landing pad on gravel or crop stubble. Avoid tall grass.

Time	Activity	Notes
At field	Run pre-flight checklist: propellers, battery, GPS, firmware	Don't skip this — three minutes of checking prevents expensive mistakes.
At field	Set up automated flight plan in mapping app	DroneDeploy: select field, set altitude (200-250 ft), overlap (75/65%), tap Fly.
Launch	Let the drone acquire GPS lock before takeoff (usually 15-30 seconds)	Wait for Ready to Fly confirmation on screen.
During flight	Monitor battery, altitude, and progress. Stay aware of surroundings.	Watch the drone periodically. Note any aircraft in the area.
After landing	Check images on SD card (random sample: sharp, well-exposed?)	If blurry or overexposed, reflying now is easier than discovering later.
Back at home	Transfer images to computer. Start processing. Create notes.txt.	Don't put this off — process the same day while the flight is fresh.

Season Summary Template

Fill this out at the end of your first season. Keep it with your drone data files. It's the document you'll reference when planning next year.

Item	Your Notes
Total fields flown	
Total flights completed	
Total flight hours	
Most useful flight (which field, which stage, why)	

Item	Your Notes
Least useful flight (which field, which stage, why)	
Management decisions informed by drone data	
Estimated value of those decisions	
Total first-year cost (equipment + operating time)	
Net first-year ROI (value minus cost)	
Biggest surprise (something you didn't expect)	
Biggest lesson learned	
Equipment upgrade needed for next year?	
Next year's priority fields and questions	

Chapter Summary

- Phase 1 (Weeks 1–4, pre-season): Order equipment, register with FAA, set up software, and practice until the flight-to-map workflow is routine. Budget 8–12 hours total.
- Phase 2 (Weeks 5–8, early season): First crop flights at V2–V4 for stand assessment, second flight at V6–V8 for baseline health. Ground-truth at least once. Make at least one management decision from the data.
- Phase 3 (Weeks 9–12, mid-season): Critical nitrogen-assessment flights at V8–V10, act on the data (side-dress, scouting, spot spray), health assessment at VT–R2. This is where the highest ROI occurs.
- Phase 4 (late season through winter): Optional late flights for documentation, collect yield data at harvest, and conduct the winter review that closes the loop between drone data and outcomes.

- The winter review is the most important work of the season: overlay drone maps with yield, calculate ROI, document what worked, and plan next year's flights.
- Your first crop flight day follows a structured sequence: check weather, charge batteries, clear SD card, pre-flight checklist, automated mission, verify images, process same day.
- RGB-only farmers are fully capable of running this 12-week plan.
- Multispectral adds NDRE-based nitrogen analysis but is not required for stand counts, weed scouting, drainage, or visual health assessment.
- Fill out the season summary template at year's end. First-year ROI for most operations ranges from $2,000–$10,000 against a $1,200–$1,500 equipment cost—if you act on what the data shows.

Chapter 11 Notes and References

1. Federal Aviation Administration. "DroneZone: UAS Registration." https://faadronezone.faa.gov/. Accessed February 2026.

2. Federal Aviation Administration. "The Recreational UAS Safety Test (TRUST)." https://www.faa.gov/uas/recreational_flyers/knowledge_test_updates. Accessed February 2026.

3. DroneDeploy. "Getting Started: Your First Mapping Flight." https://www.dronedeploy.com/. Accessed February 2026.

4. Pix4D. "Pix4Dcapture: Flight Planning for Mapping." https://www.pix4d.com/product/pix4dcapture. Accessed February 2026.

5. DJI. "DJI Mini 4 Pro: Setup and Flight Guide." https://www.dji.com/mini-4-pro. Accessed February 2026.

6. Chapters 2, 8, 9, 11, 12, 13, 14, 15, and 19 of this book provide the detailed guidance referenced in this action plan.

Note: This 12-week timeline is calibrated for a central Ohio and western Indiana growing season (late April planting through mid-September grain fill). Adjust timing for your specific region and planting date. The week numbers are flexible – what matters is matching flights to growth stages, not to specific calendar dates.

PART FOUR — Practical Applications — From Maps to Management Decisions

Chapters 12–16

You can fly the drone. You can read the maps. Now what? Part IV connects aerial data to the management decisions that actually make or save money. Chapter 12 tackles nitrogen management—using NDRE maps to put the right rate in the right place at the right time. Chapter 13 covers early-season scouting: stand counts, emergence assessment, and the replant decision. Chapter 14 builds a systematic mid-season monitoring program timed to critical growth stages. Chapter 15 addresses disease and pest detection—what drones can realistically identify and the ground-truthing protocol that prevents expensive mistakes. Chapter 16 takes the next step into variable-rate prescriptions, bridging the gap between seeing variability on a map and applying inputs with precision.

Chapter 12: Nitrogen Management—Seeing Deficiency Before It Costs You

Nitrogen is the single largest variable input cost in corn production and the one where precision pays the biggest dividend. This chapter connects everything you've learned—reflectance, indices, maps, ground-truthing—to the decision that matters most: putting the right amount of nitrogen in the right place at the right time.

Parts One through Three gave you the foundation: why aerial monitoring matters, how plants communicate through reflected light, which vegetation indices to use, how to fly and process maps, and how to read what those maps are showing you. Now we apply it. Part Four is about management decisions, and we're starting with nitrogen because no other input decision offers the same combination of financial impact, time sensitivity, and drone-data utility.

On a typical Western Ohio corn operation, nitrogen represents \$70–\$110 per acre in direct input cost. On 800 acres of corn, that's \$56,000–\$88,000 in nitrogen alone. University research consistently shows that 10–20% of that total is either wasted (applied where it wasn't needed) or misallocated (insufficient where the crop could have used more). Drone-based NDRE data—translated into zone-based management—can capture a meaningful portion of that waste.

This chapter walks you through the complete nitrogen management workflow: from the biology of why timing matters, through the practical process of turning NDRE maps into rate decisions, to the honest limitations and caveats that keep this technology grounded in reality.

The Nitrogen Timing Window: Why Days Matter

Corn's nitrogen demand isn't uniform across the growing season. The plant takes up relatively little nitrogen early – perhaps 20–30 pounds per acre by V6. Then, from about V8 through VT, nitrogen uptake accelerates dramatically. The plant consumes 60–70% of its total nitrogen during this six-to-eight-week window. After pollination, uptake slows and eventually stops.

This creates a critical management window: the period between V6 and V12 when two things are simultaneously true. First, the plant's nitrogen demand is ramping up steeply, so deficiency at this stage has maximum yield impact. Second, there's still time to intervene with a side-dress application that the crop can actually use. Once the crop reaches VT, the practical window for ground-based side-dress application closes—the corn is too tall for most equipment.

THE NITROGEN TIMELINE IN CORN

V4–V6: Low demand. Plant has taken up ~25 lbs N/acre.

Deficiency hard to detect even with sensors.

V6–V8: Rapid N uptake begins. Roughly 25% of total season demand occurs in this window.

V8–V10 (critical): Peak side-dress window. NDRE maps are most diagnostic here.

V8–V12: Peak side-dress window. NDRE maps are most diagnostic.

This is the sweet spot for drone-guided nitrogen decisions.

V12–VT: Demand at maximum. Side-dress still possible with

high-clearance equipment but window narrowing. Visible deficiency

symptoms may begin appearing.

VT–R1: Pollination. Ground application impractical.

Aerial/foliar rescue applications still possible but with diminishing

returns.

R2+: Too late for meaningful nitrogen intervention. Focus shifts to documentation and planning for next year.

The Visual vs. Sensor Timeline

Here's the point that makes drone-based nitrogen monitoring financially transformative: by the time nitrogen deficiency is visible to the naked eye—that pale green or yellowing that creeps up the lower leaves—yield loss is already occurring. Corn doesn't show visible nitrogen stress until the deficiency has been affecting the plant for one to three weeks. During that hidden period, the plant is already rationing nitrogen, slowing cell division, and reducing its yield potential.

NDRE detects the chlorophyll reduction that precedes visible yellowing. The red-edge band is sensitive enough to pick up a 10–15% decline in leaf chlorophyll content—a decline that's invisible to your eyes but already affecting photosynthetic capacity. In practical terms, an NDRE map at V8 can identify nitrogen-deficient zones 7–14 days before you'd notice the problem from the road or during a field walk.

Those 7–14 days are worth money. A side-dress application at V8, guided by NDRE data, catches the deficiency during the peak uptake window. The same application at V12—after you've finally noticed the yellowing—comes 2–3 weeks later, after the crop has already lost yield potential during the deficit period. Research from Purdue and the University of Nebraska–Lincoln suggests that each week of nitrogen stress during the V8–VT window costs approximately 3–8 bushels per acre in final yield, depending on severity.

Detection Method	Typical Detection Stage	Days Before Side-Dress Window Closes	Estimated Yield Already Lost
NDRE drone map	V8–V10	14–21 days	Minimal (0–2 bu/acre)
Tissue sampling (routine)	V8–V10	14–21 days	Minimal (0–2 bu/acre)

Detection Method	Typical Detection Stage	Days Before Side-Dress Window Closes	Estimated Yield Already Lost
Visual observation (careful scouting)	V10–V12	7–14 days	Moderate (3–8 bu/acre)
Obvious visual symptoms ("firing")	V12–VT	0–7 days	Significant (8–15+ bu/acre)

Notice that tissue sampling and NDRE mapping have similar detection timelines. The advantage of NDRE isn't that it detects deficiency earlier than all other methods—it's that it maps the deficiency across the entire field simultaneously. A tissue sample tells you the nitrogen status at the point where you pulled the sample. An NDRE map tells you the nitrogen status at every point in the field, revealing the spatial pattern that drives variable-rate decisions.

Turning Maps Into Prescriptions: Three Approaches

You have an NDRE map of your cornfield at V8–V10. It shows variability—some areas healthy, some areas stressed. Now what? The path from map to management depends on your equipment capabilities and comfort level. Here are three approaches, from simplest to most sophisticated.

Approach 1: The Simple Target (No VRT Required)

You don't own variable-rate equipment. Your sidedress applicator applies a single rate across the whole field. Does the NDRE map still help? Absolutely.

Look at the map and answer one question: is there a significant portion of the field that appears nitrogen-deficient? If the map shows 80% of the field with healthy NDRE values (0.40+) and 20% with low values (below 0.30), you know two things. First, your base nitrogen rate was sufficient for most of the field. Second, approximately 20% needs more. If the deficient area is contiguous (not scattered in small patches), you may be able to make a targeted pass over just that zone with a flat-rate applicator, skipping the healthy areas entirely.

Even if the deficiency is too scattered for targeted application, the map changes your whole-field rate decision. Without the map, you might side-dress the entire field at 60 lbs N/acre as planned. With the map showing 20% of the field significantly deficient, you might increase the whole-field rate to 70–75 lbs to ensure the stressed areas get enough—or you might decide that the stressed areas have a root cause other than nitrogen (drainage, compaction) and skip the rate increase entirely.

The map gives you information that improves your decision, regardless of equipment.

Approach 2: The Zone Method (Basic VRT)

If you have variable-rate capability—even basic zone-based VRT that lets you assign different rates to 3–5 management zones—NDRE maps become directly prescriptive. Here's the process:

Step 1: Classify the NDRE map into zones. Most agriculture-specific mapping software (Pix4Dfields, DroneDeploy, Sentera FieldAgent) can automatically divide your NDRE map into management zones based on value ranges. Start with three zones for simplicity:

Zone	NDRE Range	Map Color	Interpretation	Recommended N Rate Adjustment
Zone 1 (High vigor)	Above 0.45	Dark green	Adequate nitrogen. Crop is healthy and N-sufficient.	Reduce by 10–20 lbs/acre from base rate

Zone	NDRE Range	Map Color	Interpretation	Recommended N Rate Adjustment
Zone 2 (Moderate)	0.35–0.45	Yellow-green	Moderate chlorophyll. Near-adequate but may benefit from supplemental N.	Apply base rate (no change)
Zone 3 (Low vigor)	Below 0.35	Red-orange	Below-adequate chlorophyll. Strong nitrogen response likely.	Increase by 20–40 lbs/acre above base rate

Step 2: Assign nitrogen rates. Using the zone map and the adjustments above, assign a specific nitrogen rate to each zone. If your planned base side-dress rate was 60 lbs N/acre, Zone 1 gets 40–50 lbs, Zone 2 gets 60 lbs, and Zone 3 gets 80–100 lbs. The total field application may be similar to the flat rate, but the distribution shifts nitrogen from where it's not needed to where it is.

Step 3: Generate the prescription file. Export the zone map as a shapefile or application map file compatible with your equipment. Pix4Dfields, DroneDeploy, and QGIS can all generate these files. Load the file onto your rate controller via USB drive or wireless transfer.

Step 4: Apply and verify. Run your side-dress application using the prescription. Verify that the controller is switching rates at zone boundaries by monitoring the rate display during application. Save the as-applied map for comparison with the next drone flight.

THE 3–5 DAY PLANNING WINDOW

The complete workflow – fly the field, process the NDRE map, create zones, generate the prescription file, load it onto equipment – takes 3–5 days. Plan your flight timing accordingly: fly 3–5 days before

the applicator is scheduled. Don't fly the day before and expect to

have a prescription ready by morning. Build the processing time into

your schedule.

Approach Three: Continuous Variable-Rate (Advanced VRT)

Advanced variable-rate systems can move beyond discrete zones to continuous rate variation—adjusting the nitrogen rate on the go based on a high-resolution prescription map. Instead of three or five zones with fixed rates, the controller smoothly transitions between rates across the field, following the gradient of the NDRE map.

This approach requires more sophisticated processing. The NDRE values must be calibrated against known nitrogen response data—typically from strip trials or university research—to generate an actual rate recommendation at every point rather than just a relative zone classification. Software like Pix4Dfields and DroneDeploy offer this capability, but it generally requires more experience and often benefits from agronomist input.

If you're working with a precision agriculture consultant or a co-op's precision team, they may already have the calibration data and software capability to generate continuous prescriptions from your drone maps. This is an excellent use of a service relationship—you provide the NDRE map, they provide the prescription, and the prescriptionist's expertise bridges the gap between raw data and rate recommendation.

WHICH APPROACH IS RIGHT FOR YOU?

• *No VRT equipment: Use Approach 1 (Simple Target). The map*

improves your flat-rate decision and guides targeted scouting.

• *Basic VRT (zone capable): Use Approach 2 (Zone Method). Start*

with three zones your first year; expand to five as you gain confidence.

• *Advanced VRT + precision support: Use Approach 3 (Continuous VRT). Work with an agronomist to calibrate NDRE values to rate recommendations.*

• *First year with any system: Start with Approach 1 or 2 regardless of equipment. Build experience and confidence before going continuous.*

What the Research Shows: The Economics of Precision Nitrogen

The financial case for drone-guided nitrogen management has been studied extensively. Here's a summary of the evidence and a calculator for estimating the potential return on your operation.

Research Summary

University trials across the Corn Belt have consistently demonstrated that spatially variable nitrogen management—applying different rates to different parts of the field based on crop need—can reduce total nitrogen use by 8–20% while maintaining or improving yields compared to uniform application. The savings come from two sources: reducing over-application in areas that don't need as much nitrogen, and increasing application in areas that were previously under-supplied and dragging down average yield.

Source	Location	N Savings	Yield Impact	Notes
Purdue University (2019–2022)	Indiana	10–18%	Maintained or +3–5 bu/acre in deficient zones	Multi-year trials on variable soils
University of Illinois (2020–2023)	Illinois	8–15%	Maintained yield; reduced environmental loss	Combined NDRE + soil sampling approach

Source	Location	N Savings	Yield Impact	Notes
Ohio State University (2021–2024)	Ohio	12–20%	+2–7 bu/acre in low zones; maintained high zones	On-farm trials with cooperating farmers
University of Nebraska (2018–2022)	Nebraska	15–25%	Maintained yield under VRT; slight increase	Irrigated and dryland trials
USDA-ARS collaborative (2022)	Multiple	10–15%	Maintained yield across most sites	Meta-analysis of precision N trials

The consistent finding across studies: you can reduce total nitrogen application by 10–15% on average while maintaining yield. In some cases, yield actually increases because the previously under-supplied zones get more nitrogen. The environmental benefit is real too—less excess nitrogen means less nitrate leaching into groundwater and tile drainage, and less nitrous oxide emissions. But the financial case stands on its own.

Savings Calculator

Plug in your own numbers to estimate the potential return:

YOUR OPERATION'S POTENTIAL NITROGEN SAVINGS

A. Total corn acres: ______ acres

B. Current total N rate (all applications): ______ lbs/acre

C. Current N cost per pound: $______ /lb

D. Total N spend: A x B x C = $______

Conservative savings (10%): D x 0.10 = $______

Moderate savings (15%): D x 0.15 = $______

Aggressive savings (20%): D x 0.20 = $______

Example: 600 acres x 180 lbs/acre x $0.50/lb = $54,000 total N spend

10% savings = $5,400 | 15% savings = $8,100 | 20% savings = $10,800

Compare your potential savings to the cost of the drone and processing system (from Chapter 7's break-even analysis). For most operations above 300 corn acres, even the conservative 10% savings exceeds the annual cost of multispectral ownership. At 600 corn acres, the savings from nitrogen alone are 2–4 times the equipment cost.

Decision Guide: Is the Stress I'm Seeing Actually Nitrogen?

This is the most important question in this chapter, and it's the one that separates profitable nitrogen management from expensive mistakes. NDRE maps show chlorophyll content. Low chlorophyll is frequently caused by nitrogen deficiency—but not always. Before you spend money on supplemental nitrogen, you need to rule out other causes.

Use this diagnostic flowchart when your NDRE map shows a low-chlorophyll zone:

Step 1: Check the Pattern

Does the low-NDRE area follow a recognizable pattern? If it's aligned with equipment passes (parallel or perpendicular to rows), the cause is likely equipment-related, not nitrogen. If it follows soil type boundaries, soil characteristics (drainage, organic matter, pH) may be the primary driver. If it's in low spots or along drainage paths, waterlogging may be restricting nitrogen uptake even when nitrogen is physically present in the soil.

Step 2: Check the Timing

Did the stress appear suddenly or develop gradually? Nitrogen deficiency develops gradually over 1–3 weeks as the plant depletes available soil N. If a zone went from healthy to severely stressed between two flights taken a week apart, suspect something other than nitrogen—disease, herbicide injury, or an acute event like ponding.

Step 3: Ground-Truth the Plants

What do the plants look like up close? Nitrogen deficiency in corn produces a specific visual pattern: yellowing starts at the leaf tip and moves down the midrib, appearing first on the lower (oldest) leaves. The plant cannibalizes nitrogen from old leaves to feed new growth. If the yellowing is in the upper canopy, on leaf margins, or interveinal (between the veins), it's probably not nitrogen – consider sulfur, potassium, magnesium, or disease.

Step 4: Consider the Soil

Is the nitrogen actually missing, or is the plant unable to access it? In compacted soils, saturated soils, or soils with root-restricting layers, the nitrogen may be present but the roots can't reach it. Adding more nitrogen to a compacted zone won't help—you'll spend money on an input that sits in the soil unused. The case study in Chapter 10 (the Darke County compaction wedge) is a perfect example.

Step 5: Pull a Tissue Sample (If Still Uncertain)

If ground-truthing leaves you uncertain, pull a tissue sample from the stressed area and a comparison sample from a healthy area. Send both to your lab with a request for nitrogen (total N or nitrate-N) analysis. Turnaround is typically 3–5 business days. The tissue test gives you a definitive answer on whether the plant itself is nitrogen-deficient, regardless of what's happening in the soil.

Observation	Likely Nitrogen	Probably Not Nitrogen	Needs Further Investigation
Pattern follows soil type / drainage			

Observation	Likely Nitrogen	Probably Not Nitrogen	Needs Further Investigation
Pattern follows equipment passes			
Gradual onset over 2–3 weeks			
Sudden onset (< one week)			
Lower-leaf yellowing, tip to midrib			
Upper-leaf symptoms, interveinal			
Uniform across soil types			
Only in low/wet spots	(waterlogging limiting N uptake?)		
Compacted zone (probe test confirms)	(root restriction, not N supply)		
Tissue test confirms low N			

THE HONEST CAVEAT

Not every yellow spot needs more nitrogen. In our experience, roughly 60–70% of low-NDRE zones in mid-season corn are genuinely nitrogen-responsive. The other 30–40% are driven by drainage, compaction, disease, or other root-zone issues that nitrogen won't fix. The ground-truthing and diagnostic steps above are what separate profitable precision management from expensive misdirection.

In-Season Rescue Applications: When You Spot Deficiency Late

What if you fly at VT and discover nitrogen deficiency you didn't catch earlier? The side-dress window has closed – the corn is too tall for ground equipment. Is there anything you can do?

The answer is a qualified yes. Late-season nitrogen rescue applications can salvage some yield, but the returns diminish rapidly with growth stage, and the logistics are more complex.

Aerial Application (Crop Dusting)

Aerial applicators can apply liquid nitrogen (UAN 28% or 32%) or foliar nitrogen products over standing corn at any height. This is the most common rescue approach. The nitrogen is applied to the canopy and taken up through both leaf absorption and root uptake after it washes into the soil.

Realistic expectations: Research from Purdue and Ohio State shows that aerial nitrogen applied at VT–R2 typically recovers 30–50% of the yield that would otherwise be lost to deficiency. If the deficiency would have cost you 20 bushels per acre, a rescue application might save 6–10 bushels. At $4.50 corn, that's $27–$45 per acre in recovered yield against $15–$25 per acre in application and product cost. The math works on enough acres to be worthwhile, but it's not a full recovery.

High-Clearance Side-Dress

Some operations have access to high-clearance applicators (Hagie, Rogator, or similar) that can navigate V12–VT corn. If you identify deficiency at V10–V12 and can schedule a high-clearance rig within a week, ground-applied side-dress nitrogen is more efficient than aerial. The window is narrow, but the uptake efficiency is better because the product is placed in the root zone rather than on the canopy.

When Rescue Isn't Worth It

If the deficient area is small (under 5 acres), the economics of a rescue application rarely work. The fixed cost of calling an aerial applicator or mobilizing high-clearance equipment exceeds the recovered yield value on small acreages. Document the problem, plan for next season, and focus your current-season efforts on areas where the return justifies the cost.

Similarly, if the deficiency is caused by something other than nitrogen supply (compaction, waterlogging, root disease), rescue nitrogen won't help regardless of application method. This is why the diagnostic flowchart earlier in this chapter matters – it prevents you from spending rescue money on a non-nitrogen problem.

Limitations, Caveats, and Calibration

Drone-based nitrogen management is powerful, but it's not magic. Being honest about the limitations helps you set realistic expectations and avoid the kind of disappointment that leads people to abandon useful technology.

NDRE Is Relative, Not Absolute

An NDRE value of 0.35 doesn't tell you the plant needs exactly 40 more pounds of nitrogen. It tells you the plant has lower chlorophyll than a plant reading 0.45 in the same field. The map shows you relative variability—where the crop is doing better or worse within a field. Translating that relative signal into absolute rate recommendations requires calibration against your specific soils, hybrids, and management practices.

Single Flights Are Snapshots

A single NDRE map captures the crop's nitrogen status at one moment in time. Soil nitrogen is dynamic—it mineralizes, leaches, denitrifies, and gets taken up by the crop continuously. A field that looks nitrogen-deficient after two weeks of heavy rain may look fine a week later after the soil dries and nitrogen becomes available again. Where possible, base decisions on the trend across two flights rather than a single snapshot.

Calibration Takes Time

The generic zone thresholds we provided (NDRE above 0.45 = reduce, 0.35–0.45 = maintain, below 0.35 = increase) are starting points based on regional research. Your fields will have their own optimal ranges based on soil type, organic matter, hybrid selection, and management history. Expect to spend two to three seasons refining your zone thresholds as you learn what NDRE values mean on your specific acres.

The most effective calibration tool is the strip trial: within a field you plan to fly, leave several strips at your standard nitrogen rate while varying the rate on adjacent strips. When you fly, compare the NDRE values across rates. Over two or three seasons, you'll develop field-specific response curves that make your prescriptions dramatically more accurate.

Other Factors Affect NDRE

Hybrid genetics, planting date, population, sulfur status, and even the previous crop in the rotation can influence NDRE values independently of nitrogen. A field that followed a high-residue corn crop may show lower NDRE values in spots where residue tied up nitrogen during decomposition – a real nitrogen issue, but one driven by rotation rather than application rate.

These confounding factors are why ground-truthing remains essential even after years of experience. The map raises the question; boots on the ground answer it.

Integration Guide: Getting Prescriptions Into Your Equipment

The last-mile problem in precision nitrogen management is getting the prescription file from your computer into the rate controller on your equipment. Here's a quick guide for the most common systems.

Rate Controller / Display	File Format	Transfer Method	Notes
John Deere (Gen 4 display) via Operations Center	Shapefile (.shp)	USB or wireless via JD Operations Center	Import prescription in Operations Center; sync to display wirelessly or via USB
Raven Viper 4 / Viper Pro	Shapefile (.shp) or AgX format	USB drive	Load shapefile directly to controller; field matching required
Ag Leader InCommand / Versa	Shapefile (.shp)	USB drive or Ag Leader SMS Advanced	SMS creates and transfers prescriptions; USB also works
Trimble / CFX-750 / FMX	Shapefile (.shp)	USB drive via Farm Works	Farm Works or Connected Farm imports prescriptions
Case IH AFS / Pro 700	Shapefile (.shp)	USB or AFS Connect via AFS Connect platform	Similar workflow to JD Operations Center
Custom applicator	Shapefile (.shp) or PDF zone map	Email or USB	Provide the shapefile; their equipment handles the rest

The shapefile (.shp) format is the universal standard. If your mapping software can export a shapefile with rate attributes, virtually every VRT controller can read it—though the import process varies by manufacturer. If you're working with a custom applicator, simply providing the shapefile and letting them handle the equipment integration is often the simplest path.

Chapter Summary

- Nitrogen is the highest-impact input decision for drone-guided management. The V8–V12 side-dress window is the sweet spot where NDRE data has the greatest financial value.
- NDRE detects nitrogen deficiency 7–14 days before visible symptoms appear—a window that translates directly into saved yield. Each week of unaddressed nitrogen stress during V8–VT costs an estimated 3–8 bushels per acre.
- Three approaches for turning NDRE maps into nitrogen decisions: simple target adjustment (no VRT needed), zone-based prescriptions (basic VRT, 3–5 zones), and continuous variable-rate prescriptions (advanced VRT with agronomist support).
- University research across the Corn Belt consistently shows 10–15% nitrogen savings with spatially variable management, with yield maintained or improved. On a 600-acre corn operation, that's $5,400–$8,100 in annual savings at current nitrogen prices.
- Not every low-NDRE zone is nitrogen-deficient. An estimated 30–40% of low-chlorophyll areas are caused by drainage, compaction, disease, or root-zone issues. The five-step diagnostic process (pattern, timing, plant symptoms, soil condition, tissue test) prevents costly misdirection.
- In-season rescue applications (aerial or high-clearance) can recover 30–50% of yield loss from late-detected deficiency, but the economics only work on larger affected areas (5+ acres) where the cause is confirmed as nitrogen supply.
- NDRE values are relative within a field, not absolute rate recommendations. Calibrating zone thresholds to your specific operation takes 2–3 seasons of experience and benefits from strip trials.
- The shapefile format is the universal standard for VRT prescriptions.

- Plan a 3–5 day window between flying and application to allow for processing, zone creation, and file transfer.

Chapter 12 Notes and References

1. Abendroth, L.J., Elmore, R.W., Boyer, M.J., and Marlay, S.K. "Corn Growth and Development." Iowa State University Extension, PMR 1009, 2011.

2. Scharf, P.C., Kitchen, N.R., Sudduth, K.A., et al. "Field-Scale Variability in Optimal Nitrogen Fertilizer Rate for Corn." Agronomy Journal, 97(2), 452–461, 2005.

3. Thompson, L. and Puntel, L. "Optimize your Nitrogen Investment with Precision Agriculture Technologies." University of Nebraska–Lincoln CropWatch, November 2021.

4. Sawyer, J.E. and Nafziger, E.D. "Regional Approach to Making Nitrogen Fertilizer Rate Decisions for Corn." Proceedings of the North Central Extension-Industry Soil Fertility Conference, 2005.

5. Ohio State University Extension. "Tri-State Fertilizer Recommendations for Corn, Soybeans, Wheat, and Alfalfa." Bulletin 974, revised 2020.

6. Hatfield, J.L., Gitelson, A.A., Schepers, J.S., and Walthall, C.L. "Application of Spectral Remote Sensing for Agronomic Decisions." Agronomy Journal, 100(S3), 117–131, 2008.

7. Solari, F., Shanahan, J., Ferguson, R., Schepers, J., and Gitelson, A. "Active Sensor Reflectance Measurements of Corn Nitrogen Status and Yield Potential." Agronomy Journal, 100(3), 571–579, 2008.

8. USDA Economic Research Service. "Fertilizer Use and Price." https://www.ers.usda.gov/data-products/fertilizer-use-and-price/. Accessed February 2026.

9. Purdue University Extension. "Sidedress Nitrogen Application in Corn." https://ag.purdue.edu/. Accessed February 2026.

10. University of Illinois Extension. "Nitrogen Management for Corn: Timing, Rate, and Source." Crop Sciences. https://extension.illinois.edu/. Accessed February 2026.

Note: Nitrogen rates, savings estimates, and yield impacts cited in

this chapter are based on published university research and regional data for the central Corn Belt. Actual results will vary based on your specific soils, hybrids, weather, and management history. The savings calculator provides estimates, not guarantees. Work with your agronomist or extension specialist to calibrate recommendations to your operation.

Chapter 13: Scouting, Stand Counts, and Early Season Decisions

The first three weeks after planting set the trajectory for the entire season. A drone and an RGB camera are all you need to make the most expensive early-season decisions—replant, spray, wait—with data instead of guesswork.

Chapter 12 made the case for multispectral-driven nitrogen management—the highest-dollar application of drone technology in row crop agriculture. This chapter makes a different case: that some of the most valuable drone applications require nothing more than the standard RGB camera on a consumer drone, and that these applications happen early in the season when the stakes are highest and the intervention windows shortest.

Stand counts. Emergence uniformity. Weed scouting. Drainage mapping. These are the bread-and-butter early-season uses that deliver immediate, tangible value on every flight. They don't require vegetation indices. They don't require multispectral cameras. They require your eyes, applied from a better vantage point, with the ability to see the entire field at once.

For farmers in their first year of drone use, this chapter describes the flights that will deliver your fastest return on investment. These are the applications that make the $1,200–$1,500 equipment cost look like a bargain before the Fourth of July.

Stand Assessment: Counting What You Actually Planted

Why Stand Counts Matter

Your planter monitor says you put 34,000 seeds per acre in the ground. But how many of those seeds actually produced a living plant? The gap between what the monitor reported and what's actually growing determines your yield potential for the entire season. A field planted at 34,000 but standing at 30,000 has already lost 12% of its potential population—and that 4,000-plant deficit happened invisibly, underground, during the two weeks between planting and emergence.

Traditionally, stand counts are done on foot: walk into the field, count plants in a measured section of row, multiply to estimate the full population. The problem is coverage. A typical ground-based stand count samples 0.001% of the field – a few row segments near accessible edges. If the emergence problem is in the middle of the field, you'll never find it. If it's concentrated on one soil type that you didn't happen to walk through, you'll miss it.

A drone at 150–200 feet sees the entire field. Every row, every skip, every gap, every double. In a 15-minute flight, you get complete coverage that would take a full day on foot – and you get it with photographic precision that can be re-examined, measured, and compared for years.

Timing Your Stand Count Flight

The optimal window: V2–V4 for corn, V2–V3 for soybeans. At this stage, plants are large enough to be individually visible from 150–200 feet (2–6 inches tall with visible leaves) but small enough that each plant is distinct against the soil background. Earlier than V2, some seeds may still be emerging and you'll undercount. Later than V4 in corn (V3 in soybeans), the canopy begins to overlap and individual plants become harder to distinguish.

Flight altitude: 100–150 feet for stand counts. This is lower than the 250-foot recommendation for general monitoring. The lower altitude gives you sub-inch resolution – critical for distinguishing individual plants, especially in soybeans where plants are smaller. You'll cover less ground per battery (25–35 acres), but the detail is worth the extra battery swap.

Conditions: Any clear weather works. Stand counts rely on visual contrast between green plants and brown soil, not spectral precision.

Sunny, overcast, early morning, midday – all fine. The main thing to avoid is shadows from adjacent trees that might obscure the row pattern at field edges.

Methods for Counting

The simplest approach, and perfectly adequate for most farms. Process your images into an orthomosaic, then zoom in to full resolution and manually count plants in several representative sample areas. Use a fixed-size sample frame – say, a 1/1000th-acre section of row – and count every plant within it. Repeat at 10–15 randomly selected locations across the field. Average the counts and multiply to estimate per-acre population.

MANUAL STAND COUNT METHOD

For 30-inch rows:

1/1000th acre = 17 feet 5 inches of row

Count all plants in that 17'5" section

Multiply your count x 1,000 = estimated plants per acre

Example: Count 32 plants in a 17'5" section -> 32 x 1,000 = 32,000 plants/acre

Repeat at 10–15 locations across the field. Average the results.

This takes 20–30 minutes at your computer – no walking involved.

The advantage over ground counts: you can place your sample areas anywhere in the field, including the middle, without driving through standing crop.

Several drone mapping platforms now offer automated or semi-automated plant counting from high-resolution RGB imagery. The software uses image recognition to identify individual plants and generate a population heat map showing plant density across the entire field.

DroneDeploy: Offers a “Plant Count” feature in their agriculture package. It processes your orthomosaic and returns a population estimate by zone. Accuracy varies but is generally within 5–10% of manual counts under good conditions.

Pix4Dfields: Includes a vegetation counting module that works with RGB and multispectral imagery. Best results at V2–V3 when plant/soil contrast is highest.

Stand Count Apps (various): Several standalone apps and AI-based services accept drone images and return plant counts. Quality varies; test any new tool against manual counts before trusting it for replant decisions.

Software-assisted counting is faster than manual counting for large areas and provides field-wide coverage rather than sample-based estimates. The trade-off is accuracy – algorithms can miscount in areas with heavy residue, variable soil color, or overlapping plants. For critical decisions (replant), we recommend verifying software counts with manual spot-checks in at least three areas.

Sometimes you don't need an exact number—you need a quick answer to the question "Is it bad enough to replant?" In that case, simply reviewing the orthomosaic at full resolution gives you a rapid assessment. Scroll across the field and look for problem areas: sections where plants are clearly missing or where emergence is dramatically uneven. If the field looks uniformly populated with only scattered gaps, you probably don't need a precise count. If you see large zones of missing plants, that's your trigger for a more detailed count in those areas.

This visual triage takes five minutes and is how most experienced drone users start every stand assessment—scan the whole field first, then count only where it matters.

The Replant Decision: Data Over Gut Feel

The replant decision is one of the most stressful calls in farming. You're weighing the cost of replanting (seed, fuel, labor, delayed maturity) against the yield loss from a reduced stand. Getting it wrong in either direction is expensive: replanting when you shouldn't wastes $40–$75 per acre and pushes maturity later; failing to replant a severely damaged stand locks in yield loss for the entire season.

Drone imagery dramatically improves this decision by answering the two questions that matter most: how many plants are actually standing, and how are they distributed across the field?

The Key Variables

Remaining stand population. What's the actual plant count in the damaged area? This is where your drone stand count becomes critical. University replant guidelines are based on final stand – you need a number, not an impression.

Uniformity of the remaining stand. A field averaging 28,000 plants per acre with evenly distributed plants will yield very differently from one averaging 28,000 but with alternating sections of 34,000 and 22,000. The drone map reveals this distribution – something a traditional walk-and-count method misses entirely.

Percentage of field affected. Is the damage on 10 acres or 80 acres of a 160-acre field? Replanting 10 acres may be worthwhile even at moderate population loss; replanting 80 acres is a much larger commitment. The drone map lets you measure the damaged area precisely.

Calendar date and replant window. How late is it? Replanted corn in Western Ohio sees yield penalties of approximately 1–1.5 bushels per acre per day after May 20. By early June, the penalty is 2+ bushels per day. A field with 26,000 plants on May 15 is a different calculation than the same field on June 5.

Replant Decision Calculator

Remaining Stand (corn)	Expected Yield (% of full)	Replant Recommended?	Key Consideration
Above 30,000	95–100%	No	Yield loss minimal; replant cost exceeds benefit

Remaining Stand (corn)	Expected Yield (% of full)	Replant Recommended?	Key Consideration
26,000–30,000	85–95%	Marginal – depends on date and uniformity	Calculate replant cost vs. yield loss; later dates favor keeping
22,000–26,000	75–85%	Likely yes if before June 1	Significant yield loss; replant often justified if window open
18,000–22,000	60–75%	Yes if replant window allows	Major stand loss; replant unless very late planting date
Below 18,000	Below 60%	Yes – or consider prevented planting	Severe loss; evaluate insurance options alongside replant

These thresholds are guidelines for corn at 30-inch row spacing in the central Corn Belt. Soybean replant thresholds are lower because soybeans compensate for stand loss through branching; stands above 80,000–100,000 plants per acre rarely justify replanting.

REPLANT ECONOMICS: A QUICK EXAMPLE

Scenario: Late-May hailstorm damages 40 acres of a 160-acre cornfield. Drone imagery shows remaining stand of 24,000 plants/acre in the damaged area. Rest of field is fine at 33,000.

Option A: Keep the stand.

Expected yield at 24,000: ~80% of full potential = ~168 bu/acre (vs. 210 target)

Loss: 42 bu/acre x 40 acres x $4.50 = $7,560 in yield reduction

Option B: Replant the 40 acres.

Replant cost: $55/acre x 40 acres = $2,200 (seed + fuel + labor)

Late-plant penalty: ~8 bu/acre (May 28 replant) x 40 acres x $4.50 = $1,440

Total replant cost: $2,200 + $1,440 = $3,640

Decision: Replant saves ~$3,920 ($7,560 – $3,640). Replant justified.

> Without the drone, this farmer would have estimated the damage from the road. Many farmers underestimate or overestimate by 30–50%. Accurate acreage and population data turns this from a gamble into arithmetic.

Emergence Variability: The Invisible Yield Robber

Stand count tells you how many plants emerged. Emergence uniformity tells you how evenly they emerged—and that distinction matters more than most farmers realize.

Research from Purdue's Dr. Bob Nielsen and others has demonstrated that uneven emergence—where some plants emerge 2–3 days or more after their neighbors—costs yield even when the final stand count is at target. Late-emerging plants become "weed plants" that are permanently dominated by their earlier neighbors. They receive less light, produce smaller ears, and contribute less to yield. Studies show that corn fields with highly variable emergence can yield 5–10% less than uniform stands at the same population.

From the ground, emergence variability is nearly impossible to assess across a whole field. You can see the plant in front of you and the few plants around it, but you can't see the field-wide pattern. From 150 feet up, the pattern is unmistakable: areas where plants are uniformly sized and spaced versus areas where a mix of plant sizes indicates staggered emergence.

What to Look For

Uniform green rows with consistent plant size: Good emergence. Plants came up together and are developing at the same rate. No action needed.

Rows with mixed plant sizes (tall and short plants interspersed):

Variable emergence. The short plants emerged later and will underperform. The cause is usually planting depth inconsistency, seed-to-soil contact issues, or variable soil temperature at planting. Check planter downforce settings and closing wheel adjustment.

Zones of smaller plants corresponding to soil type or topography:

Soil-driven emergence delay. Cold, wet soil types (like Pewamo silty clay loam in Western Ohio) emerge later than well-drained types. Not fixable in-season, but valuable information for next year's planting sequence or variable-rate population decisions.

Band or strip of delayed emergence across the field: Likely a planting-date effect – the planter crossed that area when conditions were different (wetter, colder, crusted). Or a depth-adjustment issue where the planter changed depth mid-field.

Linking Emergence Maps to Action

Observation	Likely Cause	This-Season Action	Next-Season Action
Variable plant size within rows, random distribution	Planting depth inconsistency	None (can't fix in-season)	Check and calibrate planter depth control; inspect closing wheels
Variable plant size, follows soil type boundaries	Soil temperature/moisture variability	None in-season	Adjust planting sequence (plant warm soils first); consider variable-rate population

Observation	Likely Cause	This-Season Action	Next-Season Action
Entire rows delayed vs. adjacent rows	Individual row unit issue (depth, downforce)	None in-season	Inspect specific row unit; check downforce settings
Crusted area with poor emergence after rain	Soil surface crusting	Rotary hoe if still early enough	Consider residue management or row cleaner adjustment
Scattered missing plants, no pattern	Seed quality, insect damage, or disease	Scout for insects; check seed lot records	Adjust seed treatment; consider soil insecticide if rootworm area

Early Weed Scouting: The Pre-Canopy Advantage

The window between crop emergence and canopy closure is a golden opportunity for weed scouting from the air. Once the crop canopy closes—typically V8–V10 in corn, V5–R1 in soybeans—weeds are hidden beneath the crop and invisible from above. But before closure, the contrast between crop rows (straight, regular, evenly spaced) and weed patches (irregular, between rows, often clustered) makes weed identification straightforward from a drone.

What You Can See With RGB

Weed patches between rows. At V4–V6, crop rows are visible as parallel green lines with bare or residue-covered soil between them. Weed patches appear as irregular green masses between the rows that break the clean row pattern. You can't identify the species from 200 feet (a waterhemp plant looks a lot like a pigweed plant from altitude), but you can see where weeds are and where they aren't.

Herbicide-resistant escapes. After a post-emergence herbicide application, resistant weed patches stand out dramatically. The treated crop and susceptible weeds look normal; the resistant patch is a green island that survived the application. From the ground, you might see a few plants near the field edge. From the air, you see the full extent – and it's often larger than expected.

Herbicide performance evaluation. Fly the field 7–10 days after a post-emergence application. The RGB imagery shows where the herbicide worked (clean rows, no green between them) and where it didn't (persistent green patches between rows). This is actionable intelligence: the areas where the herbicide failed need a follow-up treatment before the canopy closes and makes access difficult.

Mapping Weed Patches for Spot-Spray

One of the highest-ROI applications of drone scouting is creating spot-spray maps. Instead of treating the entire field with a post-emergence herbicide because you saw weeds in one area, you fly the field, identify the actual weed patches, and treat only those areas.

The process is straightforward:

13. Fly the field at V4–V6 (or 7–10 days after a herbicide application). Use standard RGB at 200–250 feet.
14. Review the orthomosaic and mark weed patches. In your mapping software, draw boundaries around the areas with visible weed pressure. Most software lets you export these boundaries as shapefiles or KML files.
15. Calculate the total weed area. How many acres actually need treatment? On a typical well-managed field, the answer is often 10–30% of the total acreage.
16. Treat only the marked areas. If you have GPS guidance on your sprayer, load the weed-patch boundaries and spray only within them. If not, use the map as a visual reference – mark the patches with flags during ground-truthing and direct the applicator to those zones.

THE SPOT-SPRAY SAVINGS

Field size: 160 acres

Post-emergence herbicide cost: $16/acre (product + application)

Whole-field application: 160 acres x $16 = $2,560

Spot-spray (25% of field has weeds): 40 acres x $16 = $640

Savings: $1,920 on a single application, single field.

Multiply across your operation and across multiple applications per season. On a 600-acre operation, switching even one whole-field spray to spot-spray puts $5,000–$8,000 back in your pocket annually. The drone pays for itself on herbicide savings alone.

Tracking Resistant Weed Patches Over Time

Drone imagery creates a permanent, georeferenced record of where herbicide-resistant weeds are on your farm. Fly the same fields year after year, map the resistant patches, and you build a multi-year picture of how the resistance is spreading (or being contained).

This is invaluable for long-term weed management strategy. If a waterhemp patch is expanding year over year, your current herbicide program isn't containing it. If it's stable or shrinking, your management is working. Without consistent aerial documentation, this assessment relies on memory and anecdote—poor substitutes for data.

Drainage Assessment: The Flight You Didn't Know You Needed

Of all the applications in this chapter, drainage assessment may deliver the highest long-term value per flight. A single RGB flight after a significant rain event reveals your field's drainage infrastructure with a clarity that no other method—short of systematically probing the entire field—can match.

What Rain Reveals

Fly within 24–48 hours after a 1.5-inch or greater rain event. What you'll see in the RGB imagery:

Functioning tile lines. Active tile lines show as strips of darker-green, taller crop running through areas of lighter-green or stressed crop. The tile is removing excess water, allowing the roots in that strip to access oxygen and nutrients while the surrounding undrained area is still saturated. The contrast is often dramatic and visible for 3–5 days after a rain.

Broken or clogged tiles. A tile line that's working on one end and not on the other reveals itself as a healthy strip that suddenly transitions to a stressed zone. The point of transition is approximately where the break or blockage occurred. This single observation can save you thousands of dollars in diagnostic and repair costs by directing the backhoe operator to the right spot.

Surface drainage patterns. Water flows along paths of least resistance. From 250 feet, you can trace these paths across your entire field – seeing how water moves from the high points to the low points, where it concentrates, and where it exits the field. This is the information you need for waterway design, grading decisions, and surface inlet placement.

Ponding areas. Low spots where water stands for hours or days after a rain show up as dark, reflective patches or as zones of stunted, stressed crop. Mapping these consistently across multiple rain events reveals which low spots are chronic problems (worth tiling) and which are occasional nuisances (tolerable).

Turning Drainage Maps Into Investment Decisions

Tile drainage is one of the largest capital investments a Midwest farmer makes – $800–$1,500 per acre for pattern tile, and $500–$1,000 for targeted repairs or additions. Drone imagery helps you direct that investment where it will have the greatest impact.

Drone Observation	Probable Drainage Issue	Estimated Repair Cost	Estimated Annual Yield Recovery
Healthy strip ends abruptly; stress downstream	Broken or crushed tile line	$500–$2,000 (spot repair)	$500–$1,500/year on affected area
Chronic wet zone, no tile visible	Untiled low area needing drainage	$800–$1,500/acre (new tile)	$200–$500/year per acre drained

Drone Observation	Probable Drainage Issue	Estimated Repair Cost	Estimated Annual Yield Recovery
Water standing on surface, adjacent tile working	Surface inlet needed	\$500–\$1,200 per inlet	\$300–\$800/year on ponded area
Multiple tile lines visible, one missing	Missing lateral or unmapped gap in system	\$1,000–\$3,000 (new lateral)	\$500–\$1,200/year on affected strip

The numbers above are approximate, but the pattern is consistent: targeted drainage improvements guided by aerial observation typically pay for themselves within 2–4 years through recovered yield. Without the aerial view, these problems either go undiagnosed (losing yield indefinitely) or are addressed inefficiently (repairing the wrong section, tiling areas that don't need it).

THE BEST DRAINAGE FLIGHT YOU'LL EVER MAKE

Fly every field you own within 24–48 hours after the next 2-inch rain. Save the images permanently. This single set of flights will reveal more about your drainage infrastructure than any other data source available. Tile lines you didn't know existed. Breaks you've been losing yield to for years. Surface flow patterns that explain variability you've never been able to diagnose.

Total cost: zero (you already own the drone). Total time: a few hours. Potential value: thousands of dollars in targeted drainage investment over the coming years.

The Early-Season Flight Checklist

Here's a consolidated checklist for your first flights of the season, typically conducted between late May and mid-June for corn and

soybeans in the Midwest.

Timing	Flight Purpose	What to Look For	Camera / Altitude	Potential Decisions
V2–V4 (7–14 days after planting)	Stand count	Plant population, skip patterns, doubles, emergence gaps	RGB / 100–150 ft	Replant decision, planter adjustment
V3–V5	Emergence uniformity	Plant size variability, delayed zones, soil-type patterns	RGB / 150–200 ft	Planter calibration, next-year planning
V4–V6 (pre-canopy closure)	Weed scouting	Weed patches between rows, herbicide escape areas	RGB / 200–250 ft	Spot-spray map, follow-up treatment decisions
24–48 hrs after 1.5"+ rain	Drainage assessment	Tile lines, breaks, ponding, surface flow patterns	RGB / 250–300 ft	Tile repair, drainage investment planning
After storm event	Damage documentation	Hail swath, wind damage, flooding extent	RGB / 200–250 ft	Insurance claim, replant evaluation

Notice that every flight on this list uses an RGB camera. Every one is achievable with a $1,200 drone. And every one addresses a decision worth hundreds to thousands of dollars. This is the 80% solution from Chapter 6 in action—the highest-value applications of drone technology, delivered by the simplest equipment.

Chapter Summary

- Stand counts from drone imagery provide field-wide population data in 15 minutes—complete coverage that would take a full day on foot. Fly at V2–V4 at 100–150 feet for best results.
- Three counting methods: manual sample count from the orthomosaic (simplest), software-assisted automated counting (fastest), and visual triage without counting (quickest for go/no-go decisions).
- The replant decision improves dramatically with drone data. Accurate population counts and precise damaged-area measurements turn a gut-feel gamble into a calculable economic comparison. The worked example shows $3,920 in savings from one correct replant call on 40 acres.
- Emergence uniformity matters independently of population. Fields with highly variable emergence can yield 5–10% less than uniform stands at the same count. Drone imagery reveals uniformity patterns invisible from the ground.
- Pre-canopy weed scouting enables spot-spray strategies that can save $1,500–$8,000 per season by treating only the 10–30% of the field that actually has weed pressure, instead of blanket-treating the entire field.
- Tracking resistant weed patches year over year with georeferenced drone maps builds a data record that guides long-term herbicide strategy – far more reliable than memory.
- Post-rain drainage flights are among the highest-value flights you'll make. A single flight reveals functioning tiles, broken tiles, surface drainage patterns, and chronic ponding areas – guiding drainage investments that pay for themselves within 2–4 years.
- Every application in this chapter uses RGB only. Every one is achievable with the $1,200–$1,500 entry-level drone setup from Chapter 2. These are the first flights that deliver positive ROI.

Chapter 13 Notes and References

1. Nielsen, R.L. "Effect of Plant Spacing Variability on Corn Grain Yield." Purdue University, Agronomy Department. AGRY-91-01, revised 2001.

2. Nielsen, R.L. "Estimating Yield and Dollar Returns from Corn Replanting." Purdue University, Corny News Network, May 2023. https://www.agry.purdue.edu/ext/corn/news/

3. Nafziger, E.D. "Corn Replant Decisions." University of Illinois, Department of Crop Sciences, 2022.

4. Lindsey, A.J. and Thomison, P.R. "Corn Replant Considerations." Ohio State University Extension, C.O.R.N. Newsletter, 2023.

5. Licht, M. and Archontoulis, S.V. "Influence of Corn Seeding Rate, Spacing, and Emergence on Grain Yield." Iowa State University Extension, 2021.

6. Heap, I. "The International Herbicide-Resistant Weed Database." www.weedscience.org. Accessed February 2026.

7. USDA NRCS. "Conservation Practice Standard: Subsurface Drain (Code 606)." National Handbook of Conservation Practices.

8. Zucker, L.A. and Brown, L.C. "Agricultural Drainage: Water Quality Impacts and Subsurface Drainage Studies in the Midwest." Ohio State University Extension Bulletin 871, 1998.

9. Ohio State University Extension. "Replant Decisions for Corn." https://extension.osu.edu/. Accessed February 2026.

10. Purdue University Extension. "Soybean Replant Decisions." https://ag.purdue.edu/. Accessed February 2026.

Note: Replant thresholds, yield estimates, and cost figures referenced in this chapter are approximate guidelines for Midwest corn and soybean production. Actual replant economics depend on your specific hybrid, planting date, crop insurance coverage, and local conditions. Work with your seed dealer, insurance agent, and extension specialist for operation-specific recommendations.

Chapter 14: In-Season Crop Health Monitoring—The Sweet Spot

You don't need to fly every day. You don't need to fly every week. You need to fly at the right moments—the growth stages where what you see from the air changes what you do on the ground. Three or four well-timed flights tell you more than thirty flights at random.

Chapters 12 and 13 covered the two bookend applications of drone monitoring: nitrogen management (which demands multispectral precision) and early-season scouting (which thrives on simple RGB). This chapter sits in the middle—literally and figuratively. It's about the ongoing, in-season health monitoring that most farmers will do with their drones most of the time: periodic flights through the growing season that track how the crop is developing, flag emerging problems, and build the comparative data that makes every subsequent season's maps more informative.

This is the sweet spot of agricultural drone use. Not the flashiest application. Not the most technically demanding. But the one that delivers the most consistent, year-over-year value by keeping you connected to what's happening across every acre you farm—not just the acres you can see from the road.

The core philosophy is simple: fly a few times per season at key growth stages, compare each flight to the last, and investigate anything that changed. The discipline of periodic monitoring catches problems that a single flight would miss and builds the institutional knowledge that transforms good farmers into great ones.

The "Few Flights Per Season" Approach

There's a temptation, when you first get a drone, to fly everything all the time. The technology is exciting, the images are compelling, and the urge to check on your fields weekly is real. Resist it—at least at first.

The law of diminishing returns applies aggressively to flight frequency. Your first flight of the season (stand assessment at V4) delivers enormous value. Your second flight (nitrogen status at V10) delivers strong value. Your third flight (disease and stress check at R2) delivers good value. A fourth flight (late-season assessment at R5) delivers moderate value. A fifth, sixth, and seventh flight at the same general growth windows? You're mostly confirming what you already know.

The exception is event-driven flights – after storms, heavy rain, or when something looks wrong from the road. Those are always worthwhile because they respond to new information. But for routine monitoring, three to four strategic flights per crop per season is the sweet spot where time invested and information gained are in balance.

THE ROI CURVE OF FLIGHT FREQUENCY

Flight 1 (early season): High ROI. Stand counts, replant decisions, planter diagnostics. Value: $$$

Flight 2 (mid-season): High ROI. Nitrogen management, pre-canopy health baseline. Value: $$$

Flight 3 (reproductive): Moderate-high ROI. Disease detection, pollination stress, fungicide decisions. Value: $$

Flight 4 (late season): Moderate ROI. Yield prediction, harvest planning, end-of-season documentation. Value: $

Flights 5+: Diminishing ROI unless event-driven. Value: marginal for routine monitoring.

Corn Monitoring: Flight-by-Flight Guide

Here's a detailed guide to what you should be looking for at each flight timing in corn. This builds on the seasonal calendar from Chapter 8 by adding specific observation checklists, red-flag indicators, and guidance on distinguishing normal variability from problem variability.

Flight 1: V4–V6 (Late May / Early June)

Primary purpose: Stand assessment, emergence uniformity, early stress identification, planter performance evaluation.

Camera: RGB is sufficient. Fly at 150–200 feet for stand-count resolution.

Best index: NDVI or SAVI if multispectral available (bare soil still visible; SAVI reduces soil interference). RGB orthomosaic for visual assessment.

Observation	Normal or Problem?	Action If Problem
Uniform green rows, consistent plant size	Normal	None – record as baseline
Scattered individual missing plants (< 5%)	Normal	Record; no action unless concentrated
Rows or zones with > 10% missing plants	Problem	Evaluate replant (Chapter 13)
Alternating good/poor rows at regular intervals	Problem (planter)	Identify which row unit; inspect and repair
Uneven plant sizes within rows	Problem (emergence)	Check planting depth; inspect closing wheels
Green patches between rows	Problem (weeds)	Map for spot-spray; evaluate herbicide timing
Yellow or pale plants in low spots	Monitor	Could be temporary waterlogging; re-check at Flight 2

RED FLAGS AT V4–V6

Population below 24,000 on more than 10% of the field -> Immediate replant evaluation

Systematic row failures across the entire field -> Major planter issue; stop planting other fields with same equipment until diagnosed

Standing water visible more than 48 hours after the last rain -> Drainage failure; document and plan repair

Flight 2: V10–V12 (Late June / Early July)

Primary purpose: Nitrogen status assessment, pre-tassel health baseline, confirmation of earlier-detected issues, equipment problem validation.

Camera: Multispectral strongly preferred for NDRE nitrogen assessment. RGB still useful for visual scouting and pattern identification.

Best index: NDRE for nitrogen status (canopy is closing; NDVI beginning to saturate). NDVI still useful as a secondary reference.

Observation	Normal or Problem?	Action If Problem
Uniform NDRE values (0.40–0.50) across most of field	Normal	Record as baseline; compare at Flight 3
Zones of lower NDRE (< 0.35) covering > 5 acres	Problem (likely N)	Ground-truth; tissue sample; side-dress decision (Chapter 12)
NDRE gradient following soil type boundaries	Monitor	May be soil-driven; compare to yield history before treating
Perpendicular bands visible in NDVI/NDRE	Problem (applicator)	Confirm spreader/sprayer issue; check calibration records
Pale green in low areas, dark green on slopes	Monitor (drainage)	Likely wet soil limiting N uptake; drainage issue, not N rate
Uniform canopy but scattered off-color plants	Monitor	Could be early disease; mark for closer inspection at Flight 3

RED FLAGS AT V10–V12

NDRE below 0.30 on more than 15% of corn acres -> Severe nitrogen issue; immediate side-dress if window still open

Sudden appearance of stress pattern not present at Flight 1 -> Investigate urgently; new problem developing (possible herbicide injury, disease, or root issue)

Entire field showing lower NDRE than expected for growth stage -> Possible systemic issue (N application failure, wrong product, mis-calibrated equipment); check records immediately

Flight 3: VT–R2 (Late July / Early August)

Primary purpose: Pollination stress assessment, disease onset detection, fungicide decision support, canopy health under reproductive-stage demand.

Camera: Multispectral preferred for NDRE (detecting stress in closed canopy). RGB provides disease scouting value through visual color differences.

Best index: NDRE is the primary tool here. NDVI is fully saturated at this canopy density and provides limited diagnostic value. RGB orthomosaic for visual disease scouting.

Observation	Normal or Problem?	Action If Problem
Uniform dark-green canopy in NDRE	Normal	Crop is healthy through pollination
Zones of declining NDRE compared to Flight 2	Problem	Ground-truth: disease? drought stress? late N burndown?
Irregular patches of lighter canopy (RGB)	Problem (likely disease)	Walk the patches; identify disease; evaluate fungicide

Observation	Normal or Problem?	Action If Problem
Stress concentrated on field edges or low spots	Problem (environment)	Edge: spray drift or headland compaction. Low spots: drainage
Uniform slight decline across entire field	Monitor (normal senescence)	Normal as crop shifts energy to grain fill; compare to previous years
Dead or lodged areas	Problem (severe)	Weather damage, rootworm, or stalk rot; document for insurance

THE FUNGICIDE DECISION AT VT–R2

Flight 3 timing coincides with the typical fungicide application window for corn. Drone maps can support (but not replace) the fungicide decision:

- If NDRE maps show uniform health and no disease indicators on ground-truthing, a blanket fungicide application may not be justified unless disease pressure forecasts (Tarspotter, field history) indicate high risk.
- If NDRE maps show declining zones and ground-truthing confirms disease onset, targeted fungicide on the affected areas may save money versus whole-field treatment.
- Drone maps alone cannot identify disease species. Ground-truthing is always required before making a fungicide decision.

Flight 4: R4–R5 (Late August / September)—Optional

Primary purpose: Late-season stress documentation, yield prediction, harvest planning, end-of-season record for next year's planning.

Camera: Either RGB or multispectral. At this stage, visual differences are often apparent enough that RGB captures the meaningful variability.

Best index: NDRE for stress detection. NDVI can be useful again as senescence reduces canopy density and "unsaturates" the index.

Yield prediction patterns. Areas that have been consistently healthy across all three previous flights will likely yield well. Areas that showed stress at any point—and especially those stressed at multiple flights—will likely underperform. Comparing your Flight 4 map to your combine yield map at harvest validates (or challenges) your map interpretation skills for next season.

Stalk integrity concerns. Late-season stress, particularly from disease or severe nitrogen deficiency, weakens corn stalks. If your Flight 4 map shows areas of severe late-season stress, these zones are at higher risk for stalk lodging before harvest. Consider prioritizing these areas for early harvest.

Maturity variability. Fields or zones that dry down faster than others are visible in late-season imagery as lighter or more yellow canopy. This information helps you sequence your harvest: start with the driest areas, finish with the latest-maturing zones, and minimize drying costs.

End-of-season documentation. Your Flight 4 map becomes part of the permanent field record. Combined with Flights 1–3, you have a season-long visual history that you can overlay with yield data during winter planning. Areas that showed consistent stress across multiple flights are candidates for off-season remediation—drainage, deep tillage, pH adjustment, or management zone refinement.

Soybean Monitoring: Flight-by-Flight Guide

Soybeans have a compressed canopy development timeline compared to corn – they close canopy faster and transition to reproductive stages earlier. The monitoring schedule reflects this compressed window.

Flight 1: V3–V4 (Late May / Early June)

Purpose: Stand assessment, early stress, weed pressure evaluation before canopy closure.

Key observations: Soybean stands are harder to count from altitude than corn because plants are smaller and closer together. Focus on relative uniformity rather than exact counts unless you suspect severe stand loss. Look for zones of delayed emergence (often on heavier soils), gaps from planter issues, and weed patches between rows. Soybean canopy closes quickly – this may be your only window for pre-canopy weed scouting.

Flight 2: R2–R3 (Mid-July)

Purpose: Canopy health at peak biomass, early disease scouting (frogeye leaf spot, white mold risk zones, sudden death syndrome), iron deficiency chlorosis on calcareous soils.

Key observations: By R2–R3, the soybean canopy is fully closed. NDVI saturates quickly in soybeans – switch to NDRE if you have multispectral capability. In RGB, look for color differences: the bright yellow of iron chlorosis on high-pH soils, the patchy browning of SDS, or the off-green of frogeye. White mold risk is highest in dense, humid canopy – the areas your drone map shows as the thickest, most uniform green are paradoxically the areas most at risk.

Fungicide timing: The R3 growth stage is the standard fungicide application window for soybeans. As with corn, drone maps can guide but not replace the fungicide decision. If your map and ground-truthing confirm disease presence, targeted application is justified. If the canopy is healthy and disease forecasts are low, whole-field fungicide may not return its cost.

Flight 3: R5–R6 (August)

Purpose: Late-season disease progression, stress documentation, maturity assessment, harvest planning.

Key observations: By R5–R6, soybeans are entering seed fill and beginning to senesce. Look for areas that are declining faster than the rest of the field – these may indicate SDS, brown stem rot, white mold, or soybean cyst nematode damage. Areas with SCN damage often show as irregular patches of premature yellowing, particularly on sandy knobs or in historically infested zones. Document these areas for soil sampling (SCN egg counts) in the fall.

Maturity mapping: Different soybean varieties mature at different rates, and the same variety matures at different rates across soil types. Your drone map at R6 shows you which zones are ready for harvest and which need more time. This supports harvest sequencing to minimize field losses from shatter and maximize dry-down.

Wheat Monitoring: Flight-by-Flight Guide

Wheat monitoring in the Midwest follows a different calendar than corn and soybeans – it starts in March and wraps up by late June. For farms that include wheat in the rotation, these flights fill what would otherwise be a gap in drone activity before corn and soybean monitoring begins.

Flight 1: Spring Greenup (March / Early April)

Purpose: winter-kill assessment, stand thickness evaluation, early-spring vigor.

Key observations: After a tough winter, wheat fields can show dramatic variation in survival. Thin or dead zones from ice damage, heaving, or cold injury are immediately visible from the air as brown or bare areas within the green canopy. The percentage of the field that survived the winter drives your top-dress nitrogen rate: healthy stands get the full rate, thin stands get a reduced rate (lower yield potential means lower N demand), and dead zones may be candidates for spring replant with a different crop.

This is an RGB flight. At greenup, the contrast between living wheat (green) and dead wheat (brown) is so stark that multispectral adds minimal value. Fly at 200–250 feet for field-wide coverage.

Flight 2: Flag Leaf Through Heading (Late April / May)

Purpose: Disease pressure assessment (stripe rust, Septoria leaf blotch, powdery mildew), nitrogen status evaluation, fungicide decision support.

Key observations: This is the highest-stakes wheat flight. Foliar diseases are establishing during this window, and the fungicide application decision (typically at flag leaf emergence) has a major impact on yield. NDRE maps can identify areas of below-average chlorophyll that may indicate disease onset or nitrogen deficiency. Ground-truth any low-NDRE zones immediately – if disease is present, a timely fungicide application at flag leaf can protect the top two leaves that contribute 75% of grain fill photosynthesis.

Multispectral preferred. The wheat canopy is closed at this stage, and NDRE provides the best differentiation of healthy versus stressed tissue. RGB can still reveal visible disease symptoms (rust pustules create a color shift visible from the air on severely affected areas), but NDRE catches subtler differences.

Flight 3: Grain Fill (Late May / June)

Purpose: Late disease monitoring, harvest timing assessment, documentation for planning purposes.

Key observations: By grain fill, intervention options are limited. This flight is primarily for documentation and harvest planning. Areas with heavy disease pressure will dry down faster and may be ready for harvest earlier. Zones with thick, healthy canopy will take longer. Map these differences to plan your harvest sequence and estimate relative yield across the field.

The Real Power: Interpreting Change Between Flights

A single drone map is a snapshot. Two maps of the same field, taken at different growth stages, become a story. And the story—how the crop changed between flights—is where the deepest management insights live.

The Comparison Method

After each flight, compare the new map to the previous flight's map of the same field. There are three possible outcomes for any area of the field:

Stable and healthy. The area looked good at Flight 1 and still looks good at Flight 2. No action needed. This is the outcome you want for most of your field, and it's the outcome you'll see on 70–90% of your acres in a normal year.

New stress (wasn't there before). The area looked healthy at Flight 1 but shows reduced vigor at Flight 2. Something happened between flights. This is your highest-priority investigation target because it represents an active, developing problem. Causes could include nitrogen depletion (progressive), disease onset (often rapid), herbicide injury (sudden), equipment damage, or weather impact.

Persistent stress (was there before, still there). The area showed reduced vigor at Flight 1 and continues to show reduced vigor at Flight 2. This is a structural or chronic issue – compaction, drainage, soil type limitation. It's important to document but probably not actionable in-season. Add it to your fall and winter planning list.

THE THREE-QUESTION COMPARISON

After every flight, ask these three questions for each problem area:

1. Is this new or was it here last time?

2. Is it getting worse, stable, or improving?

3. Is there still a management window to respond?

New and getting worse with a management window open = immediate action. Persistent and stable with no management window = record for off-season planning. The comparison drives the priority.

Normal Variability vs. Problem Variability

Every field shows some variability on every flight. The question isn't whether there's variability – there always is. The question is whether the variability you see is within normal range or represents a genuine problem.

Indicator	Normal Variability	Problem Variability
NDRE range within the field (e.g., 0.40–0.50)	0.05–0.10 spread	0.15+ spread (e.g., 0.30–0.50)
Percentage of field in "low" zone	5–15% below field average	> 20% significantly below average
Pattern shape	Follows known soil or drainage boundaries	New pattern not matching any known feature
Change from previous flight	Gradual, uniform progression	Sudden appearance or rapid expansion
Correspondence with management records	Explained by hybrid, soil, or application differences	Not explained by any known management factor

As you fly the same fields season after season, your calibration of "normal" improves dramatically. By year three, you'll recognize your field's typical variability pattern as easily as you recognize the face of an old friend—and you'll spot the anomalies instantly.

Record-Keeping: Building the Multi-Year Picture

The value of drone monitoring compounds over time. Year 1 gives you a snapshot. Year 2 gives you a comparison. By Year 3, you have a trend – and trends are where the real management insights emerge.

What to Record for Each Flight

Data Point	Why It Matters	Where to Store
Date, time, weather conditions (wind, clouds)	Enables comparison; explains anomalies	Flight log (Appendix H template)
Field name and crop/growth stage	Context for interpretation	Flight log + folder naming convention
Camera type and settings	Ensures maps are comparable across dates	Flight log

Data Point	Why It Matters	Where to Store
Index values (min, max, mean per field)	Tracks season-long trends	Spreadsheet or farm management software
Problem areas identified	Builds a field history of persistent issues	Map annotations + notes
Ground-truthing results	Confirms or denies map interpretations	Ground-truth form (Appendix E)
Actions taken	Closes the loop; enables ROI calculation	Farm management records

The Winter Review Session

Block two to three hours in January or February to review the entire previous season's drone maps alongside your yield maps, application records, and financial summaries. This annual review is where you:

Validate your interpretations. Did the low-NDRE zone at V10 actually yield less than the rest of the field? If yes, your map reading was accurate. If not, investigate why – the calibration learning happens here.

Quantify the value. Did the targeted side-dress application actually save nitrogen? Did the spot-spray decision actually reduce herbicide cost? Putting real numbers on real outcomes builds confidence and justifies continued investment.

Plan next season's flights. Based on what you learned, adjust your flight schedule. Maybe the Flight 4 at R5 didn't change any decisions – skip it next year. Maybe you wish you'd flown a week earlier for side-dress – adjust the timing.

Identify off-season projects. Persistent problem areas revealed across multiple flights become the basis for fall and winter projects: drainage work, deep tillage, soil sampling, pH adjustment, management zone refinement.

THE COMPOUNDING VALUE OF MULTI-YEAR DATA

Year 1: You learn what your fields look like from above. Every flight is new information.

Year 2: You have a comparison point. Abnormalities stand out because you know what "normal" looks like.

Year 3: You have trends. You can distinguish between "bad year" problems and structural issues. Your zone thresholds are calibrated to your fields.

Year 5+: Your drone maps, yield maps, soil maps, and management records form an integrated dataset that drives every major agronomic decision. You're no longer guessing about field variability—you're managing it.

Chapter Summary

- Three to four well-timed flights per crop per season deliver the best ROI. More frequent flights yield diminishing returns unless driven by specific events.
- Corn monitoring: Flight 1 (V4–V6) for stands, Flight 2 (V10–V12) for nitrogen, Flight 3 (VT–R2) for disease and pollination stress, Flight 4 (R4–R5, optional) for late stress and harvest planning.
- Soybean monitoring: Flight 1 (V3–V4) for stands and weeds, Flight 2 (R2–R3) for canopy health and disease, Flight 3 (R5–R6) for late disease and maturity assessment.
- Wheat monitoring: Flight 1 (spring greenup) for winter kill, Flight 2 (flag leaf–heading) for disease and nitrogen, Flight 3 (grain fill) for harvest planning.
- Red-flag checklists at each growth stage identify the observations that warrant immediate ground-truthing versus those that can be monitored at the next flight.
- The real power is in comparing flights: new stress (high priority for investigation), persistent stress (structural issue for off-season remediation), and stable health (no action needed).

- Normal variability (0.05–0.10 NDRE spread, following known features) must be distinguished from problem variability (0.15+ NDRE spread, new patterns, rapid change).
- Record-keeping compounds value over time. By Year 3, multi-season comparisons reveal trends that single-year data cannot. An annual winter review session validates interpretations and plans the next season.

Chapter 14 Notes and References

1. Abendroth, L.J., Elmore, R.W., Boyer, M.J., and Marlay, S.K. "Corn Growth and Development." Iowa State University Extension, PMR 1009, 2011.

2. Pedersen, P. "Soybean Growth and Development." Iowa State University Extension, PM 1945, 2009.

3. Lindsey, L.E. and Thomison, P.R. "Ohio Agronomy Guide." Ohio State University Extension, Bulletin 472, revised 2023.

4. Wise, K. and Mueller, D. "Are Fungicides No Longer Just for Fungi? An Analysis of Foliar Fungicide Use in Corn." APS Features, 2011.

5. Mueller, D.S., Wise, K.A., et al. "Corn Yield Loss Estimates Due to Diseases in the United States and Ontario, Canada." Plant Health Progress, 17(3), 211–222, 2016.

6. University of Wisconsin–Madison. "Tarspotter: Tar Spot Disease Forecasting Tool." https://tarspotter.cals.wisc.edu/. Accessed February 2026.

7. Hatfield, J.L. and Prueger, J.H. "Value of Using Different Vegetative Indices to Quantify Agricultural Crop Characteristics." Remote Sensing, 2(2), 562–578, 2010.

8. Ohio State University Extension. "Wheat Disease Management." https://extension.osu.edu/. Accessed February 2026.

9. Dorrance, A.E. "Soybean Disease Management in Ohio." Ohio State University Extension. https://extension.osu.edu/. Accessed February 2026.

10. Roth, G.W. and Yocum, J. "Use of Remote Sensing for Monitoring Crop Condition." Pennsylvania State University Extension, 2020.

Note: Growth stage timings referenced in this chapter are approximate

for the Western Ohio and Eastern Indiana growing region. Actual timing varies with planting date, hybrid maturity, and seasonal weather. Adjust flight schedules based on your crop's actual development, not calendar dates.

Chapter 15: Disease and Pest Detection—What's Possible, What's Not

A drone can show you where your crop is sick. It cannot tell you what's making it sick. That distinction is the most important sentence in this chapter. Get it right, and drones become a powerful scouting multiplier. Get it wrong, and you'll spray for the wrong disease, miss the real problem, or waste money on a treatment that wasn't needed.

Disease and pest detection is the application where farmer expectations and drone reality diverge most dramatically. The marketing suggests that drones will spot diseases early and tell you exactly what to spray. The truth is more nuanced and more useful once you understand it.

Drones are excellent at detecting that something is wrong. They show you stress patterns—areas of the field where the crop isn't performing like its neighbors. They map the spatial extent of that stress with precision you can't achieve on foot. And they do this across the entire field in a single flight, directing your scouting time to the areas that need it most.

What drones cannot do—at least not with current consumer-grade technology—is identify the specific pathogen, insect, or condition causing the stress. A low-NDRE zone at R2 could be tar spot, gray leaf spot, nitrogen burndown, drought stress, rootworm damage, or six other things. The map tells you where to walk. Your eyes, your knowledge, and sometimes a lab test tell you what's causing it.

This chapter sets honest expectations, walks through what current technology can and cannot detect, profiles the major diseases and pests of Midwest row crops from an aerial perspective, and gives you a practical framework for integrating drone data with scouting and disease forecasting tools.

What Drones Can Detect

General Stress Patterns

Diseases and pest infestations create spatial patterns in crop canopies. These patterns reflect the biology of how the organism spreads and infects—and from 250 feet up, those patterns become diagnostic clues.

Circular or radial spread. Many foliar diseases spread outward from an initial infection point. Spores land, germinate, infect, and then produce new spores that disperse to nearby plants. The result is a roughly circular patch of affected crop that expands over time. From the air, these circular patterns are distinctive. They're different from the linear patterns created by equipment (Chapter 10), and they're different from the soil-driven zones that follow geological features.

Wind-driven streaks. Some diseases spread preferentially in the direction of prevailing winds. If spores are carried by wind, the infection pattern elongates downwind from the source. In a Midwest cornfield, this might appear as a northeast-trending streak (following summer prevailing winds) originating from a field edge, waterway, or adjacent field.

Edge-first progression. Diseases that arrive from external inoculum sources—spores blowing in from another field, for example—often show up first on the upwind field edge. If your drone map shows stress concentrated on one edge and diminishing toward the field interior, external disease pressure is a strong suspect.

Random scattered patches. Soilborne diseases (like sudden death syndrome in soybeans or common smut in corn) often create irregular patches that correspond to soil conditions – wet areas, compacted zones, or areas with high pathogen populations. The patches look random from above, but they correlate with soil factors.

Canopy Color and Density Changes

Both RGB and multispectral cameras detect the visible effects of disease on the crop canopy:

Color shifts in RGB. Severe foliar diseases reduce green leaf area, causing the canopy to appear lighter green, yellow, or in extreme cases, brown. At 200–250 feet, these color shifts are visible in a standard RGB orthomosaic. You can't see individual lesions, but you can see the cumulative effect of thousands of lesions on canopy color.

NDRE and NDVI reduction. Foliar diseases reduce chlorophyll content and disrupt cell structure, both of which lower NDRE and NDVI values. The multispectral signal doesn't tell you "this is tar spot"—it tells you "this area has lower chlorophyll than that area." The pattern (circular, streaked, edge-first) provides the diagnostic context.

Canopy gaps and thinning. Severe infestations – whether disease (stalk rot, SDS), insect (rootworm causing lodging), or wildlife (deer feeding) – create visible gaps or thinning in the canopy. These are easily detected in both RGB and multispectral imagery as darker areas (where the camera sees soil through the canopy) or as zones where the canopy height visibly drops.

What Drones Can't Do

Being clear about limitations prevents expensive mistakes. Here's what current drone technology cannot reliably accomplish for disease and pest management.

Identify Specific Diseases

A drone map cannot distinguish between tar spot and gray leaf spot, between frogeye leaf spot and Septoria brown spot, or between rootworm feeding damage and compaction-induced stress. All of these conditions reduce canopy health and produce similar spectral signatures. The map shows stress; the diagnosis requires ground-truthing.

This is not a temporary limitation waiting for better sensors. It's a fundamental constraint of remote sensing at current resolution. Individual disease lesions are millimeters in size—far smaller than the pixel resolution of any drone at practical agricultural flight altitudes. What the drone sees is the aggregate effect of many lesions on the overall canopy, which is the same visual signal produced by many different stressors.

Detect Most Diseases Early Enough to Prevent Yield Loss

This is the hardest truth in this chapter. By the time a disease reduces NDRE values enough to be reliably detected from a drone, the infection is already established and causing yield loss. The early, subclinical phase of most foliar diseases—when intervention would be most effective—produces changes too subtle for current multispectral sensors to detect from standard flight altitudes.

There are exceptions. Sudden death syndrome in soybeans produces dramatic, rapid plant death that's detectable early in the progression. Severe rust infections in wheat create visible color changes relatively quickly. But for the gradual-onset diseases that dominate Midwest row crops—tar spot, gray leaf spot, frogeye—the drone typically confirms what careful ground scouting would have found a few days earlier.

The drone's value is not in being first to detect. It's in mapping the full extent of the problem across the entire field—something ground scouting cannot do efficiently.

Replace IPM Scouting

DRONES SUPPLEMENT SCOUTING. THEY DO NOT REPLACE IT.

Integrated Pest Management (IPM) scouting involves walking fields, inspecting individual plants, identifying specific organisms, counting pest populations, and assessing damage thresholds. None of this can be done from 250 feet in the air.

The drone tells you WHERE to scout. Good scouting tells you WHAT you found and WHETHER to treat. A farmer who flies a drone but never walks the field is missing half the equation. A farmer who walks the field but doesn't fly it is missing the other half. The combination is what works.

The Practical Approach: Drones as Scouting Multipliers

The most accurate way to think about drones in disease and pest management is as a scouting efficiency tool. Instead of walking a 160-acre field trying to visually assess every area, you fly the field in 30 minutes, identify the 10–15 acres that look different, and walk only those areas. Your scouting time is focused where it matters.

The Workflow

Step 1: Fly the field at the appropriate growth stage. For disease scouting, the critical windows are VT–R2 for corn and R2–R4 for soybeans. These are the stages when foliar diseases are typically establishing and when fungicide applications have the greatest benefit.

Step 2: Review the map for new stress patterns. Compare to previous flights (Chapter 14's comparison method). Any new area of reduced NDRE or visible color change is a scouting target. Pay particular attention to patterns suggestive of disease: circular spread, edge-first progression, or scattered patches in low or humid areas.

Step 3: Ground-truth the suspicious areas. Walk the marked locations. Inspect individual plants. Look for lesions, pustules, mycelium, feeding damage, or other diagnostic signs. If you can't identify the problem, collect leaf samples for lab submission.

Step 4: Make the management decision. Based on what you found on the ground – not on what the map showed – decide whether treatment is warranted. If it is, use the drone map to determine the extent of the problem: is it 5 acres or 50? Is a targeted application sufficient, or does the distribution warrant whole-field treatment?

Step 5: Document the outcome. If you treat, fly the field again 10–14 days later to assess whether the treatment arrested the progression. This builds your library of "what disease patterns look like from the air" on your specific fields—knowledge that makes future identification faster.

Disease-by-Disease Detectability Guide

Not all diseases are equally visible from the air. This guide ranks the major diseases and pests of Midwest row crops by their drone detectability – how likely you are to see them in your maps and how useful the maps are for guiding management.

Corn Diseases

Detectability: Moderate. Tar spot reduces green leaf area and creates a general canopy darkening as the black stromata cover the leaves. In severe infections, NDRE declines measurably, and the canopy takes on a dull, grayish-green appearance visible in RGB. The challenge is that early tar spot (scattered stromata, minimal leaf death) produces changes too subtle for reliable aerial detection. By the time the drone clearly shows it, the infection is moderate to severe.

Pattern from above: Tar spot typically shows as diffuse, irregular zones of reduced canopy vigor, often beginning in low-lying, humid areas of the field where conditions favor infection. It may spread across the field over 2–3 weeks, becoming more uniform as severity increases.

Management integration: Pair drone flights with the Tarspotter app (University of Wisconsin) for forecasting. If Tarspotter indicates high risk and your drone map shows early canopy changes in the vulnerable areas, that's a strong signal for fungicide application.

Detectability: Low to moderate. Gray leaf spot causes rectangular lesions on corn leaves that reduce photosynthetic area. At low-to-moderate severity, the effect on canopy reflectance is subtle and difficult to distinguish from other stressors. At severe levels, the cumulative leaf area loss becomes visible as a general canopy decline in NDRE. Detection is more reliable in no-till and continuous-corn fields where inoculum pressure is highest.

Pattern from above: Often most severe in areas with poor air circulation – low spots, near tree lines, thick canopy areas. May appear as a gradient from field edge (worse) to field interior (less severe) if inoculum is blowing in from residue.

Detectability: Moderate. NCLB creates long, cigar-shaped lesions that, when severe, significantly reduce green leaf area. At moderate-to-severe levels, the NDRE decline is detectable. This disease typically affects the lower canopy first, which NDRE is well-suited to detect since the red-edge band penetrates canopy layers more effectively than visible light.

Detectability: Moderate to high (indirect). Rootworm larvae feed on corn roots, reducing the plant's ability to take up water and nutrients. The drone doesn't see the larvae – it sees the above-ground consequences: uneven canopy height, lodged plants ("goose-necking"), and stress patterns in areas of heavy feeding. Lodging is particularly conspicuous from the air: lodged plants lean at angles that create a chaotic, irregular canopy pattern clearly distinct from the uniform upright canopy of healthy corn.

Pattern from above: Rootworm damage often appears as irregular patches or arcs, sometimes following the previous year's corn-on-corn areas. Severe lodging may appear as a "swirled" pattern visible in RGB imagery.

Detectability: Low to moderate. Rust pustules are tiny and numerous. At low severity, the effect on reflectance is negligible. At moderate-to-severe levels, the cumulative rust coverage reduces NDRE and creates a general canopy decline. Southern rust, which can progress rapidly in warm, humid conditions, becomes detectable sooner than common rust because it builds severity faster.

Soybean Diseases

Detectability: High. SDS is the most aerially detectable soybean disease in the Midwest. It causes dramatic interveinal chlorosis and necrosis – leaves turn bright yellow between the veins and then brown and drop, while the petioles remain attached. The effect on canopy reflectance is severe and rapid. In NDRE maps, SDS patches show as sharp drops in index values. In RGB, the yellow-to-brown color change is unmistakable against the green canopy.

Pattern from above: SDS typically appears in patches corresponding to wet, compacted, or poorly drained areas – the soil conditions where the fungus thrives. Patches may be irregular but often cluster along drainage paths or in low-lying field positions.

Management value: There is no effective in-season treatment for SDS. But drone maps of SDS occurrence are extremely valuable for future management: selecting SDS-tolerant varieties for affected zones, improving drainage, and adjusting planting timing. Multi-year SDS maps identify the chronically affected areas of your fields with precision that soil sampling alone cannot match.

Detectability: Low to moderate. Frogeye creates small, circular lesions with gray centers and dark borders. Individual lesions are far too small to see from the air. At moderate-to-severe severity, the cumulative effect reduces canopy reflectance enough for NDRE detection. In Western Ohio, frogeye has become more prevalent and more resistant to strobilurin fungicides, making identification and management increasingly important.

Detectability: Moderate to high (at advanced stages). White mold kills plants from the stem outward, creating dramatic canopy gaps. In advanced infections, the white mycelial growth on stems is visible when zooming into high-resolution RGB orthomosaics. The pattern from above is distinctive: patches of dead or dying plants within the densest, most closed-canopy areas of the field—the high-humidity microenvironments where the fungus thrives.

Management value: White mold risk maps can be generated from drone canopy density data. The thickest, most vigorous areas of the field at R2–R3 (paradoxically the "healthiest" looking areas) are the highest risk for white mold because the dense canopy traps humidity. Targeting fungicide applications to these high-risk zones—rather than treating the whole field—is a cost-effective strategy that drone mapping enables.

Detectability: Moderate. SCN doesn't produce a visible above-ground symptom until damage is severe. The primary aerial indicator is patchy, stunted growth – areas where the soybean canopy is shorter and thinner than the surrounding field. At moderate-to-heavy infestation levels, NDRE maps show these areas as zones of reduced vigor, often on sandier or lighter-textured soils.

Management value: Drone maps of suspected SCN damage guide targeted soil sampling. Instead of randomly sampling across the field, you sample the specific low-vigor patches to confirm SCN presence and egg counts. This targeted approach is faster, cheaper, and more informative than grid sampling for SCN.

Wheat Diseases

Detectability: Moderate. Rust diseases create a distinctive yellow-orange discoloration on wheat leaves. At moderate severity, the color change is visible in RGB imagery as a yellowish cast to the canopy, and NDRE shows measurable decline. Stripe rust typically appears in streaks or patches aligned with prevailing winds. Leaf rust tends to be more uniform.

Detectability: Low to moderate. Septoria progresses from the lower canopy upward, making early detection from above difficult. By the time the upper canopy shows symptoms visible in NDRE, the disease has been progressing for several weeks. However, the spatial pattern – often worst in low-lying, humid areas – helps direct scouting.

Insect Pests

Detectability: Low unless severe. Japanese beetles skeletonize leaves, but the damage is distributed across individual plants rather than concentrated in patches. Moderate feeding rarely produces enough canopy-level change to detect from the air. Severe infestations on soybeans can create visible thinning in the upper canopy, but by that point the damage is obvious from the ground as well.

Detectability: Very low. Stink bugs feed on individual pods, causing damage that is invisible from the air. Drone maps are not useful for stink bug management. Traditional sweep-net scouting remains the only reliable detection method.

Detectability: Low to moderate at severe levels. Heavy soybean aphid infestations can cause canopy-level symptoms – leaf curling, honeydew accumulation, sooty mold – that are detectable at very high populations. But the economic threshold for aphid treatment (250 per plant, rising population) occurs well below the aerial-detection threshold. By the time the drone sees it, significant yield loss has already occurred. Aphid management requires traditional scouting.

Disease and Pest Detectability Reference Chart

Disease / Pest	Crop	Aerial Detectability	Best Camera	Pattern From Above	Drone Management Value
Tar spot	Corn	Moderate	MS preferred	Diffuse zones, low areas	Scouting prioritization; treatment extent mapping
Gray leaf spot	Corn	Low–Moderate	MS preferred	Gradient from edges/low spots	Scouting prioritization
Northern corn leaf blight	Corn	Moderate	MS preferred	Patches, lower-canopy onset	Scouting prioritization; fungicide targeting
Rootworm (lodging)	Corn	Moderate–High	RGB sufficient	Irregular patches, lodging swirls	Extent mapping; next-year management zones
Southern/common rust	Corn	Low–Moderate	MS preferred	Diffuse spread, often uniform	Limited; confirm with ground scouting

Disease / Pest	Crop	Aerial Detectability	Best Camera	Pattern From Above	Drone Management Value
Sudden death syndrome	Soybeans	High	RGB sufficient	Sharp-edged patches in wet areas	Multi-year risk mapping; variety selection
Frogeye leaf spot	Soybeans	Low–Moderate	MS preferred	Scattered, irregular	Scouting prioritization
White mold	Soybeans	Moderate–High	RGB sufficient	Gaps in dense canopy areas	Risk zone mapping; targeted fungicide
Soybean cyst nematode	Soybeans	Moderate	MS preferred	Patchy stunting, lighter soils	Targeted soil sampling; management zones
Stripe/leaf rust	Wheat	Moderate	Either	Streaks (stripe) or patches	Scouting + fungicide timing
Septoria leaf blotch	Wheat	Low–Moderate	MS preferred	Low/humid areas first	Scouting prioritization
Japanese beetles	Both	Low	Either	General thinning, edges	Limited; ground scouting required
Aphids (soybean)	Soybeans	Low	Either	Patchy at severe levels	Very limited; scout conventionally only

Decision Guide: When to Spray Based on Drone Data

This is the question that matters most in practical terms. You've flown the field, you see a stress pattern, and you need to decide whether to apply a fungicide or insecticide. Here's the framework:

SPRAY IF ALL THREE CONDITIONS ARE MET

1. The drone map shows a new or expanding stress pattern consistent with disease (circular, edge-first, or clustered in humid areas).

2. Ground-truthing confirms the presence of the specific disease on plant tissue (lesions, pustules, mycelium identified).

3. The crop is within the fungicide application window (VT–R2 for corn, R3 for soybeans, flag leaf for wheat) with yield potential still worth protecting.

HOLD AND MONITOR IF...

- *The drone shows stress, but ground-truthing finds no disease – the cause may be environmental (drought, nutrient, compaction)*
- *Disease is present but at very low severity and not yet at economic threshold*
- *Weather forecasts suggest conditions unfavorable for further disease development (hot, dry, windy)*
- *The affected area is very small (< 5 acres) and may not justify treatment cost*

DO NOT SPRAY BASED ON DRONE DATA ALONE IF...

- *You have not ground-truthed the affected areas (the stress might not be disease)*
- *The stress pattern matches known non-disease features (compaction, hybrid boundary, cloud shadow)*

• *The crop is past the effective fungicide window (fungicides protect against new infection; they don't cure established disease)*

• *You're considering treating based on a single flight without comparing to previous maps*

Integrating Drones With Disease Forecasting Tools

Drones are most powerful when combined with other information sources. Disease forecasting tools predict the likelihood of infection based on weather, field history, and management. Drone maps confirm whether the predicted conditions have produced actual disease on your fields. The combination is far more reliable than either source alone.

Tool	Disease	What It Provides	How Drones Complement It
Tarspotter (U. of Wisconsin)	Tar spot in corn	Daily risk rating based on weather and field inputs	Fly when Tarspotter shows high risk; confirm whether disease has appeared
Sporecaster (U. of Wisconsin)	White mold in soybeans	Apothecial development risk based on soil moisture/temp	Map canopy density to identify highest-risk zones; target treatment
Crop Protection Network models	Multiple diseases	Regional disease risk forecasts	Validate regional forecasts against field-specific drone observations
State IPM alerts (OSU, Purdue)	Multiple diseases/pests	Field reports and trap data from regional scouts	Confirm whether reported regional pests are present on your fields

Tool	Disease	What It Provides	How Drones Complement It
Your own field history	All	Which fields had disease in previous years	Compare current drone maps to previous-year patterns; same problem recurring?

The workflow is straightforward: subscribe to the relevant forecasting tools for your crops and region. When they signal elevated risk, fly your fields within the next few days. Use the drone map to confirm whether the risk has materialized on your acres and to guide your ground-truthing. The forecasting tool tells you when to worry; the drone tells you where to look; your boots on the ground tell you what to do.

Documentation for Insurance and Records

When disease or pest damage reaches levels that affect yield significantly, drone maps provide documentation that supports insurance claims and management records. A timestamped, GPS-referenced aerial image showing the extent of damage is a powerful piece of evidence for crop insurance adjusters.

For insurance documentation purposes:

Fly as soon as possible after identifying the damage. The earlier you document, the more accurately the imagery reflects the initial condition.

Save both the raw images and the processed orthomosaic. Raw images retain the original timestamps and GPS metadata. The orthomosaic provides the field-wide view.

Take ground-level photographs to accompany the aerial imagery.

Aerial maps show extent; ground photos show the specific damage. The combination is more persuasive than either alone.

Notify your insurance agent promptly. Don't wait until harvest to report potential losses. Most crop insurance policies require timely notification before you destroy evidence by harvesting.

Chapter Summary

- Drones detect that something is wrong. They cannot identify what is wrong. Maps show stress patterns; ground-truthing identifies the specific disease, pest, or condition causing the stress.
- Disease patterns visible from the air include circular spread (spore dispersal), wind-driven streaks, edge-first progression (external inoculum), and soilborne patches. These patterns narrow the diagnosis before you leave the truck.
- Sudden death syndrome in soybeans and rootworm lodging in corn are the most aerially detectable problems in Midwest row crops. Subtle foliar diseases (frogeye, gray leaf spot) are the least detectable until moderate-to-severe severity.
- By the time most foliar diseases are detectable via drone, infection is established. The drone's value is in mapping extent and guiding targeted treatment, not in replacing early-season scouting.
- Never spray based on drone data alone. The three conditions for treatment: the map shows a disease-consistent pattern, ground-truthing confirms the disease, and the crop is within the effective application window.
- Integrate drones with disease forecasting tools (Tarspotter, Sporecaster, state IPM alerts) for a system that is more reliable than any single data source.
- White mold risk in soybeans can be mapped from drone canopy density data – the densest areas are highest risk. This enables targeted fungicide rather than whole-field treatment.
- Drone maps are powerful insurance documentation. Timestamped, GPS-referenced imagery showing damage extent supports claims more effectively than written descriptions alone.

Chapter 15 Notes and References

1. Mueller, D.S., Wise, K.A., Dufault, N.S., et al. "Corn Yield Loss Estimates Due to Diseases in the United States and Ontario, Canada, from 2016 to 2019." Plant Health Progress, 21(4), 238–247, 2020.

2. Dalla Lana, F., Plewa, D.E., Phillippi, E.S., et al. "Tar Spot of Corn: A Review of Biology, Epidemiology, and Management." Plant Disease, 105(8), 2141–2154, 2021.

3. University of Wisconsin–Madison. "Tarspotter: Tar Spot Disease Forecasting Tool." https://tarspotter.cals.wisc.edu/. Accessed February 2026.

4. University of Wisconsin–Madison. "Sporecaster: White Mold Forecasting." https://ipcm.wisc.edu/apps/sporecaster/. Accessed February 2026.

5. Dorrance, A.E., Cruz, C., Mills, D., et al. "Management of SDS of Soybean." Ohio State University Extension, 2023.

6. Crop Protection Network. "Disease Loss Estimates and Forecasting Models." https://cropprotectionnetwork.org/. Accessed February 2026.

7. Mahlein, A.K. "Plant Disease Detection by Imaging Sensors – Parallels and Specific Demands for Precision Agriculture and Plant Phenotyping." Plant Disease, 100(2), 241–251, 2016.

8. Zhang, J., Huang, Y., Pu, R., et al. "Monitoring Plant Diseases and Pests Through Remote Sensing Technology: A Review." Computers and Electronics in Agriculture, 165, 104943, 2019.

9. Ohio State University Extension. "C.O.R.N. Newsletter: Pest and Disease Updates." https://agcrops.osu.edu/newsletter/corn-newsletter. Accessed February 2026.

10. Purdue University Extension. "Crop Chat and Pest & Crop Newsletter." https://extension.entm.purdue.edu/pestcrop/. Accessed February 2026.

Note: Disease detectability ratings in this chapter are based on

current consumer-grade drone and sensor technology at standard agricultural flight altitudes (200–300 feet). Advances in sensor technology, AI-based image analysis, and hyperspectral cameras may improve detectability over time. The ratings represent what is practically achievable with 2025–2026 equipment, not the theoretical limits of remote sensing.

Chapter 16: Variable-Rate Prescriptions—Taking the Next Step

The drone showed you the problem. The ground-truthing confirmed it. Now you need to fix it—not everywhere, not at the same rate, but precisely where and how much. This chapter bridges the gap between seeing variability on a map and applying the right input at the right rate to the right acre.

Everything in this book up to this point has been about generating information—flying, processing, interpreting, ground-truthing. This chapter is about acting on that information through variable-rate technology. VRT is where drone data transforms from interesting aerial photographs into input prescriptions that change what comes out of your applicator, planter, or spreader at every point in the field.

We touched on variable-rate concepts in Chapter 12 (nitrogen management) and briefly in Chapter 7 (investment decisions). This chapter goes deeper: the complete workflow from drone map to prescription file, the specific inputs where VRT delivers the clearest return, the integration challenges you'll encounter, and a first-year calibration strategy that builds accuracy over time.

If you're not using VRT equipment yet, this chapter will show you what the technology makes possible and help you decide whether the investment is worth pursuing. If you're already running VRT and looking for better prescription data, this chapter shows you how drone maps fit into—and improve—your existing precision agriculture workflow.

What Variable-Rate Technology Requires

Before diving into prescription creation, let's be clear about the prerequisites. VRT isn't just software and data—it's a system that requires all components to work together.

The Equipment Chain

A GPS-equipped applicator with rate control. The applicator—whether it's a fertilizer spreader, sprayer, or planter—must have a GPS receiver that knows where it is in the field and a rate controller that can adjust output on the go. Most modern precision equipment sold in the past decade has this capability. Older equipment may need retrofit rate controllers.

A display/controller that accepts prescription files. The in-cab display reads the prescription file and sends rate commands to the applicator as you drive across the field. John Deere Generation 4, Raven Viper 4, Ag Leader InCommand, Trimble CFX/TMX, and Case IH AFS displays all support prescription files. The specific file format varies by manufacturer—we'll cover compatibility in Chapter 18's integration guide.

A prescription file. This is the file you create from your drone data. It's essentially a map of the field divided into zones (or a continuous surface), with a specific application rate assigned to each zone or point. The format is almost always a shapefile (.shp) or a proprietary format specific to the equipment manufacturer.

Software to create the prescription. This is the bridge between your drone map and the equipment. Pix4Dfields, DroneDeploy, QGIS, and various farm management platforms (John Deere Operations Center, Ag Leader SMS, Trimble Farm Works) can all generate prescription files from zone maps.

THE VRT CHAIN: FOUR LINKS

Drone map (data) -> Software (prescription) -> Display (instructions)

-> Applicator (action)

If any link is missing, the chain breaks. A drone map without VRT-capable equipment is still useful for scouting. VRT equipment without good prescription data is applying variable rates based on incomplete information. The system delivers its full value only when all four components work together.

What Drones Add to VRT

Drone data isn't the only source of VRT prescriptions. Farmers have been creating variable-rate maps from yield data, soil sampling, and satellite imagery for decades. What drones add is resolution, timeliness, and in-season adaptability.

Data Source	Resolution	Timeliness	Cost	Best For
Grid soil sampling	2.5-acre grids (typical)	Every 3–4 years	\$7–12/acre	pH, P, K—stable soil properties
Yield maps	Sub-acre (combine width)	After harvest (retrospective)	Included with monitor	Identifying persistent zones; historical trends
Satellite imagery (Sentinel-2)	30–33 ft/pixel	Every five days (weather permitting)	Free	Broad patterns; season-long monitoring
Drone imagery (RGB)	1–3 inches/pixel	On-demand (same day)	\$0 marginal (own drone)	Stand counts, drainage, visual scouting
Drone imagery (multispectral)	1–3 inches/pixel	On-demand (same day)	\$0 marginal (own drone)	NDRE for nitrogen, in-season prescriptions

The key advantage of drone data for VRT prescriptions is the combination of high resolution and in-season timing. Soil sampling tells you what the soil looks like every few years. Yield maps tell you what happened last fall. Drones tell you what the crop looks like right now, at sub-acre resolution, in time to change what you do this season. For nitrogen side-dress decisions, this combination of resolution and timeliness is unmatched.

Creating Prescriptions From Drone Data: Step by Step

This section walks through the complete workflow for creating a variable-rate prescription from a drone map. We'll use nitrogen side-dress as the example because it's the highest-value and most common application, but the process applies to any input.

Step 1: Fly the Field at the Right Time

For nitrogen prescriptions, fly at V8–V10 – the sweet spot where NDRE is most diagnostic for nitrogen status and the side-dress window is still open. Plan to fly 3–5 days before the applicator is scheduled. This gives you time to process, create zones, generate the file, and load it onto equipment.

Step 2: Process the Images and Generate the NDRE Map

Follow the processing workflow from Chapter 9. The output you need is a calibrated NDRE index map. If using Pix4Dfields, the NDRE map is generated automatically during processing. In WebODM with multispectral data, you'll need to select the appropriate band math. Set the color scale to a fixed range (0.15–0.55 for NDRE) to ensure consistent interpretation.

Step 3: Create Management Zones

This is where the map becomes a prescription. You're dividing the continuous NDRE gradient into discrete management zones, each of which will receive a different application rate.

Start with three zones your first year. Three zones are enough to capture the meaningful variability in most fields while keeping the prescription simple enough to verify and troubleshoot. As you gain experience, you can expand to four or five zones in subsequent years.

Zone	NDRE Threshold	Classification	N Rate Strategy
Zone 1 (High vigor)	> 0.45	Nitrogen-sufficient. Crop healthy.	Reduce: apply 75–85% of base rate
Zone 2 (Moderate)	0.35–0.45	Near-adequate. Moderate chlorophyll.	Maintain: apply 100% of base rate
Zone 3 (Low vigor)	< 0.35	Nitrogen-deficient. Low chlorophyll.	Increase: apply 120–150% of base rate

Important: ground-truth Zone 3 before assigning a high rate. As Chapter 12 emphasized, not every low-NDRE zone is nitrogen-deficient. If the low-NDRE area is caused by compaction, drainage, or disease, applying extra nitrogen wastes money and may increase environmental loss. Walk Zone three and confirm nitrogen deficiency symptoms before including it in the prescription.

In Pix4Dfields: Open the NDRE map, select "Zonation," set the number of zones to three, and adjust the threshold boundaries to the values above. The software automatically generates a zone map. You can manually edit zone boundaries if you know the field well enough to override the algorithm (for example, excluding a known drainage problem from the high-rate zone).

In DroneDeploy: Use the "Annotation" or "Zone" tool to draw zone boundaries based on the NDRE color map. DroneDeploy's agriculture tier includes an automatic zone generation feature. Export the zones as shapefiles.

In QGIS (free): Load the NDRE GeoTIFF. Use the Raster Calculator to reclassify the continuous NDRE values into three classes. Convert the reclassified raster to a vector (polygon) layer. Add a "rate" column to the attribute table and assign the nitrogen rate for each zone. Export as a shapefile. This method requires more GIS skill but is completely free and gives you maximum control.

Step 4: Assign Application Rates

With your three zones defined, assign a specific application rate to each. Here's a worked example for a side-dress nitrogen prescription:

WORKED EXAMPLE: SIDE-DRESS NITROGEN PRESCRIPTION

Field: 160-acre cornfield, Darke County, Ohio

Pre-plant N applied: 80 lbs/acre (anhydrous, uniform)

Planned side-dress base rate: 60 lbs/acre (28% UAN)

NDRE flight: V9, June 27

Zone 1 (high vigor, 85 acres): 45 lbs N/acre (25% reduction).

This zone has adequate N and doesn't need the full rate.

Zone 2 (moderate, 50 acres): 60 lbs N/acre (base rate).

Near-adequate; maintain planned rate.

Zone 3 (low vigor, 25 acres): 80 lbs N/acre (33% increase).

Confirmed N-deficient via ground-truthing.

Flat-rate total: 160 acres x 60 lbs = 9,600 lbs N

VRT total: (85x45) + (50x60) + (25x80) = 3,825 + 3,000 + 2,000 =

8,825 lbs N

N saved: 775 lbs = $387 savings on this single field at

$0.50/lb

Plus: Zone 3 gets 20 lbs/acre MORE than the flat rate -> potential

yield recovery of 5–10 bu/acre on 25 acres

Step 5: Generate the Prescription File

Export the zone map with rate attributes as a shapefile. The shapefile package consists of several files (.shp, .shx, .dbf, .prj)—keep them together in the same folder. The .dbf file contains the rate values that the controller reads.

Verify the file before loading it onto equipment. Open it in your farm management software or QGIS and confirm that zone boundaries look correct and rate values are assigned properly. A misplaced decimal point or a zone boundary that extends beyond the field edge can cause equipment errors in the field.

Step 6: Load, Apply, and Verify

Transfer to equipment. Copy the shapefile set to a USB drive. Insert it into the in-cab display. Navigate to the prescription import function (varies by manufacturer—refer to Chapter 18's integration table). Match the prescription field boundary to the field in the display's field map.

Test before applying. Drive through the field with the applicator off and watch the rate display change as you cross zone boundaries. Confirm that the controller is reading the prescription correctly and switching rates at the right locations.

Apply and monitor. Run the application normally, monitoring the rate display throughout. The controller should adjust the rate smoothly as you cross boundaries. Watch for any areas where the rate doesn't change when expected – this may indicate a zone boundary error or a GPS offset.

Save the as-applied map. Every modern rate controller records what was actually applied – the as-applied map. Save this data. Comparing the prescription (what you planned) to the as-applied (what actually happened) to the yield map (what the crop produced) is how you calibrate and improve prescriptions over time.

Inputs That Work With Drone-Based VRT

Nitrogen side-dress is the showcase application, but drone data can inform VRT prescriptions for several other inputs. The value and feasibility vary by input.

Nitrogen (Side-Dress and Top-Dress)

VRT value: Highest. As detailed in Chapter 12 and the examples above, NDRE-based nitrogen prescriptions offer the clearest, most immediate financial return. The crop's nitrogen demand varies significantly within most fields, the cost of nitrogen is high enough that savings are meaningful, and the in-season timing of drone data aligns perfectly with the side-dress decision window.

Seeding Rate (Variable-Rate Population)

VRT value: Moderate to high. Drone-derived management zones can guide variable-rate seeding—planting higher populations in high-productivity zones and lower populations in low-productivity zones. The logic: high-yield zones can support more plants competing for resources, while low-yield zones are resource-limited and produce better per-plant yields at lower populations.

The challenge is that seeding rate prescriptions are created before planting, which means they rely on historical drone data (previous season's maps) and yield history rather than current in-season data. Drone maps from the previous two or three seasons, combined with yield maps, create a multi-year productivity surface that's more robust than any single data source.

Zone Productivity	Typical Population Adjustment	Rationale
High (top 25% of field)	Increase by 2,000–4,000 seeds/acre above base	Strong soils support more plants; more plants capture more yield potential
Average (middle 50%)	Base rate (no change)	Standard productivity; standard population
Low (bottom 25%)	Decrease by 2,000–4,000 seeds/acre below base	Limited resources; fewer plants produce larger ears and more stable yield

At $300 per unit of corn seed, a population reduction of 3,000 seeds per acre on 40 low-productivity acres saves roughly $360 in seed cost while potentially improving per-plant yield in those zones. Across a 600-acre operation, VR seeding typically saves $1,500–$3,000 annually in seed cost alone.

Fungicide (Targeted Application)

VRT value: Moderate. Chapter 15 discussed using drone maps to identify disease-prone areas. VRT takes this a step further: instead of spraying the entire field or manually flagging areas, you generate a prescription that applies fungicide at full rate in the high-risk zones and reduced rate (or zero) in healthy zones.

For corn fungicide at VT ($15–$25/acre for product plus application), treating only the 40–60% of the field that shows disease risk or high-value canopy worth protecting can save $1,000–$2,500 per 160-acre field compared to blanket application. The prescription is based on NDRE canopy health (healthy canopy worth protecting vs. already-stressed canopy with lower yield potential) combined with ground-truthed disease observations.

Lime, Phosphorus, and Potassium

VRT value: Low from drone data alone. Soil pH, phosphorus, and potassium levels are determined by soil chemistry, not by what's visible from the air. Grid soil sampling is the appropriate data source for these inputs. However, drone-derived management zones can complement soil sampling by helping you design smarter sampling strategies – sampling more densely in variable areas identified by the drone and less densely in uniform areas.

Where drone data adds value to P and K management is in validating soil sampling zones. If your soil-based VRT prescription says a zone needs more phosphorus, but the drone shows the crop performing well in that zone, the soil prescription may be working. If the drone shows poor performance in a zone receiving high P rates, something other than phosphorus may be limiting – prompting a deeper investigation before continuing to invest in P.

Input-by-Input VRT Summary

Nitrogen (in-season): ***** Best drone-VRT application. Clear ROI. Start here.

Seeding rate: **** Strong value using multi-year drone + yield data for zone creation.

Fungicide: *** Good value for targeted treatment. Requires ground-truth confirmation.

Lime/P/K: ** Drone supports but doesn't replace soil sampling for these inputs.

Integration Challenges: The Real-World Friction

On paper, the workflow from drone map to prescription to application is clean and logical. In practice, you'll encounter friction at several points. Knowing about these challenges in advance helps you plan around them.

File Format Compatibility

The drone mapping software exports a shapefile. Your rate controller expects a specific format. Sometimes these match perfectly. Sometimes the controller wants a shapefile with a specific column name for the rate attribute, or it needs the coordinate system in a specific projection, or it requires the zones in a proprietary format that requires conversion through the manufacturer's farm management software.

The solution: Test the pipeline before the pressure is on. In the off-season, create a dummy prescription file, load it onto your equipment, and verify that the display reads it correctly. Identify any format issues and resolve them before you're standing in a field with an applicator idling and a prescription that won't load.

Timing Pressure

The most common failure point in drone-to-VRT workflows isn't technical—it's timing. The window between "crop is at the right stage for flying" and "applicator needs to be in the field" is often tight. If rain delays your flight by three days and the applicator is scheduled for tomorrow, you're either flying at a suboptimal stage, creating the prescription under time pressure, or reverting to a flat rate.

The solution: Build flexibility into your schedule. Fly at the early end of the window (V8 rather than V10) to give yourself processing and planning time. Communicate with your applicator about the 3–5 day lead time you need. Have a flat-rate backup plan so that a delayed flight doesn't cancel the entire application.

Getting Custom Applicators On Board

If you hire a custom applicator for side-dress or fungicide, they need to accept and use your prescription file. Some custom operators embrace precision agriculture and welcome prescription files. Others view them as complications that slow down their operation.

The solution: Talk to your applicator in the off-season, not the day before the job. Provide a sample file to test on their equipment. Ask what format they prefer. Some operators are more comfortable receiving a simple zone map (PDF with zones marked and rates listed) that they can implement manually than a shapefile they need to import. Meeting them where they are—even if it means a less precise implementation—is better than perfect prescription on paper that never gets applied.

GPS Offset Between Drone and Equipment

Your drone's GPS and your applicator's GPS may not agree perfectly on where the field boundaries are. A 5–10 foot offset between the prescription map and the equipment's position is common with consumer-grade GPS. This means zone boundaries may be slightly shifted—the applicator starts switching rates a few feet before or after the drone map says the zone changes.

The solution: For zone-based prescriptions with 3–5 large zones, a 10-foot offset is negligible—the zones are many acres in size. If you move to continuous VRT with fine-resolution prescriptions, consider RTK-corrected GPS on both the drone and the equipment for sub-inch accuracy. For most farmers starting out, consumer GPS accuracy is sufficient.

First-Year Calibration Strategy

Your first VRT prescription will be an educated estimate. That's fine. The goal of Year 1 isn't perfection – it's to establish a baseline from which you calibrate in Year 2 and beyond.

The Strip Trial Approach

The most effective calibration tool is embedding strip trials within your VRT prescription. Here's how it works:

1. Within each zone, leave one or two strips at the uniform base rate instead of the zone-specific rate. These strips should be full applicator-width and at least 500 feet long—large enough to harvest separately and get a clean yield comparison.
2. At harvest, compare the yield in the VRT strips to the yield in the flat-rate check strips within the same zone.
3. If the reduced-rate zone (Zone 1) yielded the same as its flat-rate check strip, your rate reduction was appropriate. If it yielded less, you cut too much. If the increased-rate zone (Zone three) yielded more than its flat-rate check, the extra nitrogen was justified. If it yielded the same, you over-applied.

After two or three seasons of strip trials, you'll have field-specific data showing exactly how much nitrogen each zone needs for optimal economic return. Your prescriptions move from estimates to calibrated recommendations—and your ROI improves with each year of data.

The Feedback Loop

THE CONTINUOUS IMPROVEMENT CYCLE

Year 1: Fly -> Create zones -> Apply VRT with check strips ->

Compare yield at harvest

Year 2: Refine zone thresholds based on Year 1 results -> Adjust rates -> Repeat with check strips

Year 3: High confidence in zone thresholds and rates -> Drop most check strips -> Focus on refinement

Ongoing: Annual comparison of prescription -> as-applied -> yield map -> next year's prescription

This feedback loop is the fundamental mechanism of precision agriculture improvement. It's not glamorous. It's not instant. But it's the process by which drone data transitions from "interesting pictures" to "calibrated management tools." The farmers who commit to this loop for three or more years report the highest satisfaction and strongest financial returns from their precision agriculture investments.

Working With Your Co-Op or Precision Agriculture Provider

You don't have to do this alone. Many co-ops, independent agronomists, and precision agriculture service providers can help at various points in the workflow—and some can handle the entire process from drone flight to prescription delivery.

Service Level	What They Do	What You Provide	Typical Cost
Full service (flight through prescription)	Fly the field, process data, create zones, generate prescription, deliver file	Field access, application schedule, rate preferences	\$10–\$20/acre per flight
Prescription creation only	Take your processed drone map, create zones, generate prescription file	Processed NDRE map (GeoTIFF), base rates, zone preferences	\$3–\$5/acre

Service Level	What They Do	What You Provide	Typical Cost
Consulting (interpretation + guidance)	Review your maps with you, recommend zones and rates, provide agronomic context	Your maps, yield history, management records	$2–$4/acre or hourly
DIY with technical support	Answer questions, troubleshoot file format issues, review your prescription	Everything – you do the work, they provide expertise as needed	Often included in co-op relationship

The most practical path for many farmers is a hybrid approach: fly the field yourself (you know your timing needs better than anyone), process the images using your own software, and then work with your co-op's precision team or independent agronomist to create the prescription and handle the equipment integration. You bring the data; they bring the prescription-creation expertise and the relationship with the applicator.

Chapter Summary

- Variable-rate technology requires four linked components: drone data, prescription software, an in-cab display, and a rate-controlled applicator. All four must work together.
- Drone maps add high-resolution, in-season prescription data that complements (not replaces) soil sampling and yield maps. The combination is more powerful than any single source.
- The prescription workflow: fly at V8–V10 -> process NDRE map -> create three management zones -> assign rates -> export shapefile -> load onto equipment -> apply and verify. Plan a 3–5 day timeline from flight to application.
- Start with three zones (high, moderate, low NDRE) and nitrogen side-dress for your first VRT prescription. Ground-truth Zone three

before assigning a high rate—low NDRE isn't always nitrogen deficiency.

- Beyond nitrogen, drone data supports VRT for seeding rate (moderate-high value), fungicide targeting (moderate value), and soil input validation (supportive role alongside soil sampling).
- Real-world integration challenges include file format compatibility, timing pressure, custom applicator adoption, and GPS offset. Test the pipeline in the off-season before the pressure is on.
- First-year prescriptions are estimates. Embed strip trials within your VRT application to calibrate zone thresholds and rates. The continuous improvement cycle (fly -> apply -> compare yield -> refine) produces high-confidence prescriptions by Year 3.
- Co-ops and precision ag providers offer services ranging from full-service flight-to-prescription to DIY technical support. The hybrid approach – you fly, they prescribe – is practical and cost-effective for most operations.

Chapter 16 Notes and References

1. Scharf, P.C., Kitchen, N.R., Sudduth, K.A., et al. "Field-Scale Variability in Optimal Nitrogen Fertilizer Rate for Corn." Agronomy Journal, 97(2), 452–461, 2005.
2. Bullock, D.S., Boerngen, M., Tao, H., et al. "The Data-Intensive Farm Management Project: Changing Agronomic Research Through On-Farm Experimentation." Agronomy Journal, 111(6), 2681–2698, 2019.
3. Lowenberg-DeBoer, J. and Erickson, B. "Setting the Record Straight on Precision Agriculture Adoption." Agronomy Journal, 111(4), 1552–1569, 2019.
4. Pix4D. "How to Create Variable Rate Application Maps." Pix4Dfields Knowledge Base. https://support.pix4d.com/. Accessed February 2026.
5. John Deere. "Operations Center: Prescription Maps." https://www.deere.com/en/technology-products/precision-ag-technology/. Accessed February 2026.
6. Ag Leader Technology. "SMS Advanced: Prescription Creation." https://www.agleader.com/. Accessed February 2026.

7. Raven Industries. "Viper 4+: Variable Rate Application." https://ravenind.com/. Accessed February 2026.

8. USDA Economic Research Service. "Precision Agriculture in the Digital Age." Economic Information Bulletin No. 248, February 2023.

9. Thompson, L. and Puntel, L. "Optimize your Nitrogen Investment with Precision Agriculture Technologies." University of Nebraska–Lincoln CropWatch, November 2021.

10. Ohio State University Extension. "Precision Agriculture for Ohio Farmers." https://extension.osu.edu/. Accessed February 2026.

Note: Equipment compatibility, file formats, and software workflows described in this chapter are based on products and versions available as of early 2026. Manufacturer updates may change specific menu names, import procedures, or supported formats. Test your specific equipment and software combination before relying on it for in-season applications.

PART FIVE — The Bigger Picture — Making It Work on Your Farm

Chapters 17–19 + Epilogue

The technology works. The question now is how to make it fit your operation. Part V addresses the practical realities of running a drone program on a working farm. Chapter 17 helps you decide when to fly yourself and when to hire a service provider – and how to get value from either approach. Chapter 18 covers integration: connecting drone data to your yield maps, soil surveys, and existing precision agriculture workflow so nothing sits unused on a hard drive. Chapter 19 catalogs the most common mistakes new drone operators make and how to avoid them – saving you the time, money, and frustration of learning the hard way. The Epilogue looks ahead at where farm drone technology is going and what it means for your operation.

Chapter 17: Working With Service Providers—When to DIY, When to Hire

You don't have to do everything yourself. The best farmers have always known when to run their own equipment and when to call in the specialist. Drones are no different. The question isn't whether to use the technology—it's whether you're the right person to operate it.

This book has assumed, for most of its pages, that you're flying your own drone, processing your own maps, and making your own management decisions from the data. That's the path we recommend for most farmers who are interested in the technology and willing to invest the learning time. But it's not the only path.

A growing industry of drone service providers offers everything from raw flights to fully interpreted reports with prescription files ready to load onto your equipment. Some farmers find that hiring out the drone work—just as they hire a custom combine operator or a soil sampling service—is a better fit for their operation, their skills, and their time.

This chapter helps you make that decision honestly: when does DIY make sense, when does hiring make sense, what should you expect from a service provider, and how do you get the most value from the relationship?

The Make vs. Buy Decision

Like most farm decisions, this comes down to economics, timing, and personal fit. Neither option is universally better—the right choice depends on your specific operation.

DIY Makes Sense When...

You'll fly 30+ hours per year. The fixed cost of drone equipment ($1,200–$5,000 for monitoring; more for multispectral) is best spread across many flights. At 30+ hours per year—roughly 3–4 flights each on 8–10 fields—the per-flight cost of ownership drops below the per-flight cost of hiring. Below 15–20 hours per year, hiring is almost always cheaper.

You want data on your schedule. This is the strongest non-financial argument for DIY. When you spot yellowing from the road on Wednesday morning, you want to fly that field Wednesday afternoon—not schedule a service visit for next Tuesday. The side-dress window from Chapter 12 doesn't wait for a provider's calendar. Owning the drone means the data is always available when the management window demands it.

You're genuinely interested in the technology. Some farmers enjoy learning new tools. If flying a drone, processing maps, and interpreting the results sounds engaging rather than burdensome, DIY will deliver more satisfaction and faster skill development. The farmers who get the most value from drone monitoring are the ones who find it interesting enough to practice, experiment, and refine their skills.

You have time to learn. Expect 10–15 hours of learning time before you're comfortable with the full workflow: flight planning, flying, image transfer, processing, and basic map interpretation. After that, each flight-to-map cycle takes 1–2 hours plus processing time. If your schedule has room for this, DIY is rewarding. If you're already stretched thin, that learning curve may be a bridge too far.

Hiring Makes Sense When...

You need multispectral without the equipment investment.

Multispectral systems cost $5,000–$15,000+. If you want NDRE data for nitrogen management but can't justify the investment on your acreage, hiring a provider with multispectral equipment gives you the data without the capital outlay. At $8–$15 per acre for 2–3 flights per season, many operations can access multispectral data for less than the annual cost of ownership.

You want processed maps without learning software. Some farmers have no interest in photogrammetry software, GIS, or digital image processing—and that's perfectly fine. A service provider delivers finished maps: orthomosaics, index overlays, management zone recommendations. You interpret the maps and make decisions; they handle the technical production.

Your time is better spent elsewhere. During the June–August monitoring season, you're also spraying, scouting, side-dressing, managing irrigation, and dealing with the hundred other demands of a growing season. If your hourly value as a farm manager exceeds the cost of hiring a drone service, delegation makes economic sense.

You need data occasionally, not regularly. If you only want drone maps for specific situations – replant evaluation after a storm, drainage assessment after a major rain, a one-time field evaluation – hiring for individual flights is far more efficient than maintaining equipment year-round.

The Hybrid Approach

Many farmers find that the best solution is a combination: own a consumer RGB drone ($1,200–$1,500) for frequent, on-demand scouting and stand counts, and hire a multispectral service provider for 2–3 high-value flights per season (nitrogen assessment, mid-season health check). This gives you the scheduling flexibility of DIY for the most common flights and the data quality of professional multispectral for the highest-stakes decisions.

THE HYBRID MODEL

Own: DJI Mini 4 Pro or similar ($1,200–$1,500). Use for stand counts, weed scouting, drainage flights, storm damage assessment, and routine visual monitoring.

Hire: Multispectral service 2–3 times/season ($8–$15/acre).

Use for NDRE nitrogen assessment (V8–V10), mid-season health baseline (VT–R2), and prescription creation.

Total annual cost: $1,500 equipment (one-time) +

$2,000–$5,000/year for hired flights on 200–400 corn acres.

This model gives you daily access to aerial scouting for $0 marginal cost, plus professional multispectral data at the critical decision points. For operations under 500 corn acres, this is often the sweet spot.

What to Expect From Service Providers

Pricing

Service Level	What You Get	Typical Price (2026)	Best For
Flight only	Raw images on a USB drive or cloud link	$3–$5/acre	Farmers who process their own maps
Flight + processing	Orthomosaic and basic index maps (NDVI, NDRE)	$5–$10/acre	Farmers who interpret their own maps
Flight + processing + interpretation	Maps plus written report with observations and recommendations	$8–$15/acre	Farmers who want turnkey analysis
Full service (through VRT prescription)	Maps, interpretation, management zones, VRT prescription files	$12–$20/acre	Farmers who want end-to-end service

Service Level	What You Get	Typical Price (2026)	Best For
Seasonal contract (3–4 flights)	Discounted per-flight pricing for monitoring season	$15–$35/acre total	Farmers wanting consistent commitment

Pricing varies significantly by region, provider type, and field size. Co-ops and ag retailers sometimes include drone services as part of a bundled precision agriculture package. Independent providers may negotiate on price for multi-field or seasonal commitments. Always get a clear quote specifying what's included before the first flight.

Deliverables

At minimum, you should receive:

Orthomosaic map. A single, stitched aerial image of the entire field at high resolution. This is the base product. If the provider can't deliver a clean orthomosaic, find a different provider.

Vegetation index maps (if multispectral). NDVI and NDRE at minimum. The maps should be delivered with a consistent color scale and value range, ideally with a legend that makes the map interpretable without the provider standing next to you.

Digital files you can keep and use. GeoTIFF format for orthomosaics and index maps is the standard. Avoid providers who only deliver a web-based viewer that you can't download from—you need files you own and can use in your own software, overlay with yield maps, and archive for future comparison.

Better providers also deliver:

Written summary or report. A brief document identifying the key observations from each flight: problem areas, recommended ground-truthing locations, comparison to previous flights, and any agronomic recommendations.

Annotated maps. Maps with markers or labels calling out specific features—"possible nitrogen deficiency," "planter skip row seven," "drainage issue here." These annotations are where the provider's agronomic expertise adds the most value.

Prescription files (if requested). Shapefiles or equipment-specific formats for VRT applications. This is the highest-value deliverable and often the difference between a "flight service" and a "precision agriculture partner."

Turnaround Time

Deliverable	Typical Turnaround	Rush Turnaround (if available)	Notes
Raw images	Same day	Same day	Just the memory card or upload
Orthomosaic	1–3 business days	Same day or next day	Processing time dependent on acreage
Index maps + report	2–5 business days	1–2 business days	Interpretation takes human time
Prescription files	3–7 business days	2–3 business days	Zone creation and rate assignment

Turnaround matters because management windows are finite. If you need an NDRE map for a side-dress decision and the provider delivers it a week after the applicator has moved on, the data is historically interesting but operationally useless. Discuss turnaround expectations explicitly before hiring, and confirm that the provider can meet your timeline during their busiest season (June–August).

Finding Service Providers

Where to Look

Local ag retailers and co-ops. Many larger ag retailers (Nutrien Ag Solutions, Helena, Growmark affiliates, and independent co-ops) have added drone services to their precision agriculture offerings. The advantage: they already know your fields, your soil, and your management history. The potential drawback: their drone service may be an add-on rather than a core competency.

Independent agronomists with drone capabilities. Certified Crop Advisors (CCAs) and independent agronomists who've added drone technology to their consulting practice combine agronomic expertise with aerial data. This is often the highest-value provider type because the person interpreting the map is the same person who understands your rotation, your soils, and your management objectives.

Regional precision agriculture companies. Companies that specialize in precision agriculture services—variable-rate prescriptions, soil sampling, yield data analysis—increasingly include drone services. They offer the most integrated service (drone data folded into your complete precision ag program) but may be more expensive and less flexible on scheduling.

University extension. Ohio State, Purdue, and University of Illinois extension programs occasionally offer demonstration flights, training sessions, or referrals to qualified providers. They won't fly your fields commercially, but they can point you toward reputable operators in your area and help you evaluate what you're getting.

Manufacturer directories. DJI, DroneDeploy, Pix4D, and Sentera maintain directories of authorized or trained service providers. These can be useful starting points, though being listed in a directory doesn't guarantee agronomic expertise.

Evaluating a Service Provider: Questions to Ask

PROVIDER EVALUATION CHECKLIST

Credentials and Experience

- [ ] Do you hold a current FAA Part 107 remote pilot certificate?
- [ ] How many seasons have you been flying agricultural fields?
- [ ] Do you have agronomic training or a CCA/agronomist on staff?
- [ ] Can you provide references from farmers in my crop/region?

Equipment and Data

- [ ] What drone and sensor platform do you fly?
- [ ] Do you fly multispectral, and if so, what sensor?
- [ ] What software do you use for processing?
- [ ] Will I receive my data in standard formats (GeoTIFF, shapefile) that I own?

Service and Scheduling

- [ ] What is your typical turnaround time for processed maps?
- [ ] How many clients are you serving during peak season? Can you meet my timing?
- [ ] What happens if weather delays the scheduled flight – when do you reschedule?
- [ ] Can you generate VRT prescription files compatible with my equipment?

Pricing and Terms

- [ ] What is the per-acre cost, and what does it include?
- [ ] Is there a minimum acreage or field size?
- [ ] Do you offer seasonal contracts with discounted pricing?
- [ ] Do you carry liability insurance? What is the coverage?

Working Effectively With Your Provider

The value you get from a drone service provider increases dramatically when you invest in the relationship. A provider who understands your fields, your management philosophy, and your decision-making timeline delivers far better results than one who shows up, flies, and emails you a map.

Communicate What You Want to Learn

Before each flight, tell the provider what questions you're trying to answer. "Fly my field and send me a map" is a vague instruction that produces a generic deliverable. "I side-dressed nitrogen two weeks ago and I want to see if the V10 NDRE map shows improvement in the zones I treated at a higher rate" gives the provider specific context that shapes how they process, interpret, and present the data.

Good questions to ask your provider to focus on: Where are the problem areas this flight? How do they compare to the last flight? Are there any patterns that suggest a specific cause? What should I ground-truth first? Is there anything I should act on immediately?

Share Your Field History

Providers can't interpret your maps in a vacuum. Share what you know: previous crop, nitrogen rates, planter population targets, hybrid placements, known problem areas, tile maps, recent weather events. The more context the provider has, the more accurate their interpretation. A low-NDRE zone means something very different if the provider knows "that's the heavy clay ground that's always wet" versus knowing nothing about the field.

Request Raw Data

Even if you're paying for processed maps and interpretation, always request the raw data files—GeoTIFF orthomosaics and index maps that you can open in QGIS, Google Earth Pro, or your farm management software. You may not use them immediately, but they become part of your permanent field record. Two or three years from now, when you're comparing this season to previous seasons, you'll want those files.

Build a Multi-Year Relationship

A provider who's flown your fields for three seasons understands your operation's baseline—what normal variability looks like, which fields have persistent issues, and how your management decisions play out in the data. This institutional knowledge makes their interpretation better every year. Switching providers annually to chase small price differences sacrifices this learning.

The ideal provider relationship resembles the relationship you have with a good agronomist or veterinarian: built on trust, informed by shared history, and oriented toward your operation's long-term success.

Key Elements of a Service Agreement

Whether formal or informal, the agreement between you and your provider should address these points. A written agreement protects both parties and prevents the misunderstandings that damage relationships.

Element	What to Specify	Why It Matters
Scope of service	Number of flights, fields, acres, deliverables for each flight	Prevents confusion about what's included in the quoted price
Pricing and payment	Per-acre rate or seasonal flat fee; when payment is due	Avoids billing surprises; defines financial commitment
Scheduling and timing	Target growth stages; scheduling window; weather-delay protocol	Ensures flights happen when the data is actionable
Turnaround time	Agreed delivery timeline for each deliverable type	Guarantees you receive data while the management window is still open

Element	What to Specify	Why It Matters
Data ownership	You own all data (raw and processed) generated from your fields	Protects your field data; prevents provider lock-in
Data format	GeoTIFF, shapefile, PDF reports; equipment-compatible prescription files	Ensures you can use the data in your own systems
Confidentiality	Provider will not share your field data with third parties	Protects your agronomic information and management decisions
Liability and insurance	Provider carries liability insurance; limits of coverage specified	Protects you if equipment causes damage or trespass occurs
Cancellation	Conditions for canceling flights or the season contract	Provides flexibility if circumstances change

DATA OWNERSHIP IS NON-NEGOTIABLE

Your field data belongs to you. Period. Any provider agreement should explicitly state that all data generated from your fields—raw images, processed maps, index layers, zone classifications – is your property. You should be able to request all data at any time, in standard formats, without restriction.

Be cautious of providers whose platforms lock your data behind a proprietary portal that you can't download from. If you can't export

your maps as GeoTIFF files, you don't really own the data. This matters when you switch providers, upgrade your own capabilities, or simply want to archive your field history.

Side-by-Side Comparison: DIY vs. Hire

Factor	DIY (Own Drone)	Hire Service Provider
Equipment cost	$1,200–$5,000	$0 upfront; provider owns equipment (multispectral: $5,000–$15,000 if purchased)
Per-flight cost	Near $0 (batteries, storage)	$3–$20/acre per flight
Annual cost (500 acres, three flights)	$500–$1,000 (consumables)	$4,500–$15,000
Scheduling flexibility	Fly any time, any day	Provider's availability; 1–7 day lead time
Turnaround time	Same day (you process)	1–7 business days
Technical skill required	Moderate (flying, processing, interpretation)	Low (interpretation only)
Agronomic interpretation	Your own knowledge	Provider's expertise included (at higher tiers)
Data quality (multispectral)	Depends on your sensor and calibration	Professional-grade; calibrated
Learning time	10–15 hours initially; ongoing	Minimal
Multi-year data consistency	High (same equipment, same process)	Varies if provider or equipment changes

Chapter Summary

- DIY makes sense when you'll fly 30+ hours per year, want data on your own schedule, are interested in the technology, and have time to learn. The per-flight cost advantage grows with volume.
- Hiring makes sense when you need multispectral without the capital investment, prefer turnkey maps and interpretation, have limited time

for drone operations, or need data only occasionally.

- The hybrid approach—own a $1,200–$1,500 RGB drone for frequent scouting and hire multispectral services for 2–3 high-value flights per season—is the sweet spot for many operations under 500 corn acres.
- Service provider pricing ranges from $3–$5/acre (flight only) to $12–$20/acre (full service through prescription files). Seasonal contracts of 3–4 flights typically offer better value than individual bookings.
- Critical deliverables: orthomosaic maps, index maps in GeoTIFF format, written interpretation, and (if needed) prescription files. Insist on data formats you own and can use independently.
- Turnaround time matters because management windows are finite. Confirm in advance that the provider can deliver processed data within your decision timeline, especially during peak season.
- Find providers through ag retailers, co-ops, independent agronomists, precision ag companies, extension programs, and manufacturer directories. Prioritize agronomic expertise alongside technical capability.
- Data ownership is non-negotiable. Your field data belongs to you. Ensure any agreement explicitly states this and provides for data export in standard formats.
- Build multi-year relationships with providers. Their understanding of your fields improves every season, and the accumulated knowledge makes their interpretation progressively more valuable.

Chapter 17 Notes and References

1. USDA Economic Research Service. "Precision Agriculture in the Digital Age." Economic Information Bulletin No. 248, February 2023.

2. Lowenberg-DeBoer, J. and Erickson, B. "Setting the Record Straight on Precision Agriculture Adoption." Agronomy Journal, 111(4), 1552–1569, 2019.

3. American Society of Agronomy. "Certified Crop Adviser Program." https://www.certifiedcropadviser.org/. Accessed February 2026.

4. Ohio State University Extension. "Precision Agriculture Resources for Ohio Farmers." https://extension.osu.edu/. Accessed February 2026.

5. Purdue University Extension. "Precision Agriculture Services and Providers." https://ag.purdue.edu/. Accessed February 2026.

6. DJI. "Enterprise Partners and Service Providers." https://enterprise.dji.com/. Accessed February 2026.

7. DroneDeploy. "Service Provider Network." https://www.dronedeploy.com/. Accessed February 2026.

8. Pix4D. "Network of Professional Drone Service Providers." https://www.pix4d.com/. Accessed February 2026.

9. Association of Equipment Manufacturers. "Precision Agriculture: Dealer and Service Provider Landscape." AEM Industry Report, 2024.

10. Griffin, T.W., Shockley, J.M., and Mark, T.B. "Economics of Precision Farming." In: Shannon, D.K., Clay, D.E., and Kitchen, N.R. (eds.), Precision Agriculture Basics, ASA, CSSA, SSSA, 2018.

Note: Service provider pricing, availability, and offerings vary significantly by region and change frequently. The price ranges in this chapter represent typical 2026 Midwest market rates. Request current quotes from multiple providers before committing. Provider recommendations in this chapter are categories of providers to consider, not endorsements of specific companies.

Chapter 18: Integrating Drones Into Your Existing System

A drone map sitting in a folder on your computer isn't helping anyone. It becomes valuable when it's connected—to your yield maps, your soil data, your agronomist's recommendations, and the decisions you make in the cab. This chapter is about making those connections.

Parts One through Four taught you the science, the operations, the management applications, and the advanced capabilities of agricultural drone technology. Part V is about making it work in the messy, practical, time-constrained reality of farming. No one has unlimited time to manage data, learn software, or reorganize their workflow around a new tool.

This chapter addresses the integration challenge directly: how to connect drone data with the precision agriculture systems you already use, how to manage the files without drowning in data, how to make your agronomist relationship more productive with aerial insights, and how to troubleshoot the friction points that cause many farmers to abandon useful technology.

The guiding principle is simplicity. The system that gets used consistently beats the perfect system that's too complicated to maintain. We'll build integration around what you probably already have rather than asking you to adopt an entirely new workflow.

Connecting the Dots: Overlaying Your Data Layers

The single biggest upgrade you can make to your drone program isn't buying a better camera or more expensive software. It's learning to overlay your drone maps with the other data you already have—yield maps, soil survey data, as-applied maps, and planting records. The overlay is where patterns become explanations and explanations become management changes.

Yield Maps + Drone Maps

You've been collecting yield data from your combine monitor for years. Those yield maps show you what happened—which areas of the field produced more or less grain. Drone maps show you why. The overlay connects the outcome (yield) to the cause (what the crop looked like during the season).

What to look for: Areas where low yield corresponds to low mid-season NDRE confirm that the stress you saw during the growing season translated into real yield loss. This validates your map interpretation. Areas where yield is low but NDRE was normal suggest a late-season problem – disease, stalk lodging, or harvest loss – that occurred after your last drone flight. Areas where NDRE was low but yield was surprisingly good suggest the crop compensated or the stress resolved itself.

How to do it: Most farm management software (John Deere Operations Center, Climate FieldView, Ag Leader SMS, Trimble Farm Works) can display yield maps and imported drone maps side by side or overlaid on the same field. Export your drone orthomosaic or index map as a GeoTIFF and import it into the same platform that holds your yield data. The GPS coordinates align the layers automatically.

Soil Maps + Drone Maps

The USDA Web Soil Survey (websoilsurvey.nrcs.usda.gov) provides free, detailed soil type maps for every field in the country. Overlaying drone maps with soil type boundaries reveals how much of the variability you see from the air is driven by inherent soil properties versus management factors.

What to look for: If your low-NDRE zones align precisely with heavy clay soil boundaries, the stress is probably soil-driven (drainage, compaction, limited rooting depth) rather than management-driven (nitrogen rate, planter issue). This distinction matters enormously for prescription decisions – you wouldn't add nitrogen to fix a drainage problem.

How to do it: Download the soil survey shapefile from Web Soil Survey and load it into QGIS or your farm management software alongside the drone map. Google Earth Pro can also display both layers if you export the drone map as a KMZ overlay and import the soil boundaries. The free tools work fine for this.

As-Applied Maps + Drone Maps

If your sprayer, planter, or fertilizer applicator records as-applied data (and most modern equipment does), overlaying that data with drone maps closes the loop between what you applied and how the crop responded.

What to look for: If the drone map shows a perpendicular band of low NDRE across the field and the as-applied map shows a gap or reduced rate on that same pass, you've identified an application error – and you can trace it to a specific time, location, and equipment issue. If the as-applied map shows uniform application but the drone shows variable crop response, the soil or some other factor is driving the variability.

How to do it: As-applied data exports from your rate controller as a shapefile or proprietary format. Import it into the same platform where you're viewing your drone maps. The John Deere Operations Center and Climate FieldView handle this overlay natively if you're using JD equipment. For other systems, QGIS can layer any combination of shapefiles and GeoTIFFs.

THE THREE-LAYER OVERLAY

The most powerful analysis you can do requires no special software, no agronomic degree, and no precision agriculture certification. Open three layers for the same field: the drone NDRE map from mid-season, the yield map from harvest, and the soil survey map. Spend 20 minutes studying where these layers agree and where they diverge. That

20-minute exercise will teach you more about your field's variability

than any single data source can provide alone.

Data Management: Keeping It Simple

Drone monitoring generates data – lots of it. A single flight of a 160-acre field at standard resolution produces 300–800 individual images totaling 5–15 GB. Processed outputs add another 1–5 GB per field. Multiply by 8–10 fields, 3–4 flights per season, and you're looking at 200–600 GB of data annually.

Without a management system, this data quickly becomes an overwhelming pile of folders with names like "DJI_0423" and "orthomosaic_final_v2_REVISED.tif." Nobody can find anything six months later. Here's a system that works.

Folder Structure

RECOMMENDED FOLDER STRUCTURE

Farm_Drone_Data/

2026/

North_80/

2026-05-28_V4_stand_count/

raw_images/

processed/

notes.txt

2026-06-27_V10_nitrogen/

2026-07-22_VT_disease/

South_160/

...

```
2025/
...
```

Format: YYYY-MM-DD_GrowthStage_Purpose. This convention sorts chronologically by default, tells you the crop stage at a glance, and describes why you flew. "2026-06-27_V10_nitrogen" is immediately understandable a year from now. "DJI_flight_27" is not.

What to Keep, What to Delete

Data Type	Keep?	Storage Size	Notes
Raw images	Keep for 1 season, then delete	5–15 GB per flight	Needed only if you want to reprocess; processed outputs contain the useful information
Orthomosaic (GeoTIFF)	Keep permanently	200 MB–1 GB per field	Your primary archive; the field photo you'll reference for years
Index maps (NDVI, NDRE)	Keep permanently	100–500 MB per field	Your analytical archive; enables multi-year comparison
Prescription files (shapefiles)	Keep permanently	< 10 MB per field	Tiny files; valuable for comparing prescriptions to outcomes
Processing project files	Delete after confirmed good	1–5 GB per processing project	Software working files; only needed for reprocessing

Data Type	Keep?	Storage Size	Notes
Flight logs	Keep permanently	< 1 MB each	Tiny; useful for records, insurance, regulatory compliance

Following this strategy, your permanent archive for one field over one season is roughly 1–3 GB: the orthomosaics and index maps from 3–4 flights. That's manageable. Over 10 years, 10 fields at 3 GB per field per year totals about 300 GB – one external hard drive.

Backup Strategy

Drone data is a farm asset – it has value that grows over time as your multi-year archive enables increasingly powerful comparisons. Protect it the way you'd protect any farm record.

The 3-2-1 rule: Keep three copies of important data, on two different types of media, with one copy stored off-site. For most farmers, this means: working copy on your computer, backup copy on an external hard drive stored in a different building, and (optionally) a cloud backup through a service like Backblaze, iDrive, or Google Drive. The external drive is the critical minimum. Cloud is a bonus.

Schedule a backup at the end of each flight day – copy the processed outputs to the external drive while the flight is fresh in your mind. Once a year (during your winter review from Chapter 11), verify the backup is complete and the drives are healthy.

Working With Your Agronomist: Better Conversations

If you work with an agronomist—whether through a co-op, as an independent consultant, or through your seed company—drone maps transform the relationship from periodic field walks to data-driven conversations.

Before the Field Visit

Share your most recent drone maps with your agronomist 24–48 hours before a scheduled field visit. Email the orthomosaic and index map as PDFs or provide a link to the files. Include a brief note identifying what you're seeing and what questions you have.

This simple step changes the dynamics of the visit. Instead of arriving at the field with no context and spending the first hour walking around looking for problems, your agronomist arrives having already studied the map, identified the likely problem areas, and formed preliminary hypotheses. The field visit becomes targeted – walk directly to the spots the map flagged, confirm or revise the hypotheses, and spend the time on diagnosis and decisions rather than discovery.

During the Conversation

Drone maps create a shared visual language between you and your agronomist. Instead of "the north end of the field looks rough," you can say "the NDRE map shows a 15-acre zone in the northeast quadrant with values 0.15 below the field average, and it corresponds to the Blount silty clay loam on the soil map." That's a different conversation—more precise, more productive, and more likely to result in a specific recommendation.

Ask your agronomist to ground-truth the problem areas identified by the map and explain what they find in the context of the aerial data. A good agronomist will connect the plant symptoms they see on the ground with the spatial patterns visible from the air, building your interpretive skills alongside their own assessment.

Tracking Recommendations vs. Outcomes

One of the most valuable but underused applications of drone data is holding recommendations accountable. When your agronomist recommends a specific rate, product, or treatment, the drone and yield data let you evaluate whether it worked.

Did the side-dress prescription recover yield in the low-NDRE zones? Did the fungicide application protect the canopy in the treated area compared to untreated? Did the variable-rate population actually optimize yield across zones? These questions can be answered by comparing the recommendation to the drone maps and yield data. Over time, this evidence-based approach strengthens both your agronomist's advice and your own decision-making.

Equipment Compatibility: Getting Data Into the Cab

The last-mile challenge in drone integration is moving data from your computer into the rate controller, GPS display, or farm management platform in the tractor cab. Chapter 12 and Chapter 16 covered prescription file transfer for VRT. This section covers the broader compatibility picture.

Common File Formats

Format	Extension	What It Contains	Used For
Shapefile	.shp (+.shx, .dbf, .prj)	Vector polygons with attribute data	Prescription zones, field boundaries, scouting points
GeoTIFF	.tif	Georeferenced raster image	Orthomosaics, index maps, elevation models
KMZ/KML	.kmz / .kml	Google Earth-compatible geographic data	Simple overlays for visualization; not for VRT
PDF map	.pdf	Printable map image	Reports, visual reference, sharing with non-technical users
CSV	.csv	Tabular data with coordinates	Scouting points, sampling locations, simple zone data

Format	Extension	What It Contains	Used For
ISO-XML	.xml	ISOBUS-compatibl e task data	Some European and newer equipment; AGCO and others

Platform-Specific Integration

John Deere's ecosystem is the most common in the central Corn Belt. Import drone-derived prescription files through the Operations Center website. Create a "prescription" within Operations Center by importing your shapefile and assigning it to a field. The prescription syncs to your Gen four display wirelessly (if connected) or via USB export. You can also import GeoTIFF orthomosaics as field imagery layers for visual reference in the display.

Case IH uses the AFS Connect platform for data management. Import shapefiles through AFS Connect or the AFS Pro 1200 display via USB. The workflow is similar to John Deere's: upload the prescription to the platform, assign to a field, sync to the display. GeoTIFF import for background imagery is supported on newer displays.

AGCO's Fuse platform supports shapefile import for prescriptions. The system also accepts ISO-XML task files, which some drone processing software can export directly. The Fuse platform's cloud integration allows wireless sync between web-based management and the in-cab display.

If your rate controller is an aftermarket unit (Raven Viper four, Ag Leader InCommand/Versa, Trimble FMX/TMX), these systems universally accept shapefiles via USB drive. The import process varies slightly by manufacturer, but the common pattern is: insert USB, navigate to "Import Prescription," select the shapefile, match the rate column to the controller's rate variable, and apply. Refer to the manufacturer's manual for specific menu navigation.

Troubleshooting Common Integration Problems

Problem	Likely Cause	Fix
Display won't read the prescription file	Wrong file format or missing component files (.shx, .dbf, .prj)	Verify all four shapefile components are in the same folder; check format requirements
Prescription loads but rates are wrong	Rate column mismatched or units wrong (lbs vs. gal)	Verify the attribute column name matches what the controller expects; check unit settings
Zone boundaries are offset from field features	Coordinate system mismatch or GPS accuracy difference	Re-export shapefile in WGS84 (most common); accept minor offsets for zone-based VRT
File is too large to load	Too many polygons or excessive detail in zone boundaries	Simplify the zone map (fewer, larger zones); reduce polygon vertex count
Controller switches rates too slowly	Zone boundaries too close together for equipment response time	Merge small zones into larger ones; minimum zone size should match equipment transition distance
Orthomosaic won't display in farm software	Unsupported projection, too large, or wrong format	Re-export as GeoTIFF in WGS84; reduce resolution if file exceeds import limits

Sample Workflow: Flight to Prescription to Application

Here's the complete integrated workflow for the most common high-value scenario: a nitrogen side-dress prescription based on a drone NDRE map, applied through VRT equipment, and verified against yield data.

Day	Activity	Tools	Output
Day 1	Fly field at V9 (NDRE)	Drone + multispectral camera	SD card with 400+ multispectral images
Day 1–2	Transfer images to computer; start processing	Pix4Dfields / WebODM	NDRE index map (GeoTIFF)
Day 2–3	Create 3 management zones from NDRE map	Pix4Dfields / QGIS	Zone map (shapefile)
Day 3	Ground-truth Zone 3 (low NDRE); confirm N deficiency	Truck + tissue sample kit	Confirmation that Zone 3 is nitrogen-responsive
Day 3–4	Assign N rates to zones; generate prescription shapefile	Pix4Dfields / QGIS	Prescription shapefile with rate attributes
Day 4	Load prescription onto rate controller via USB	USB drive + in-cab display	Prescription active in equipment
Day 4–5	Apply side-dress nitrogen using VRT prescription	Applicator + rate controller	As-applied map saved to controller
Day 14–21	Fly field again (V12–VT); compare NDRE to pre-treatment map	Drone + multispectral camera	Post-treatment NDRE map
Harvest	Collect yield map; overlay with NDRE maps and prescription	Combine monitor + farm software	Complete prescripti on-to-yield comparison

Day	Activity	Tools	Output
Winter	Review: Did the prescription improve yield in Zone 3? Refine for next year.	Computer + farm software	Updated zone thresholds and rate recommendations

This five-day workflow from flight to application is achievable with practice. The first time you do it will take longer as you learn each step. By the second or third iteration, the process becomes routine – you'll know your software, your equipment, and your fields well enough that the workflow feels natural rather than forced.

DON'T LET PERFECT BE THE ENEMY OF GOOD

If this workflow feels overwhelming, simplify. Skip the VRT prescription entirely and use the NDRE map to make a better flat-rate decision. Or use the map to identify where to ground-truth and let the rest of your season proceed normally. Any use of the drone data is better than no use. The full integrated workflow is the goal you grow into over two or three seasons, not the minimum requirement for getting value.

Chapter Summary

- Overlaying drone maps with yield maps, soil survey data, and as-applied records is the single most valuable integration step. The three-layer overlay (NDRE + yield + soil) reveals whether variability is management-driven or inherent.
- Use a simple, consistent folder structure: Year / Field / Date_Stage_Purpose. Keep orthomosaics and index maps permanently; delete raw images after one season; back up to an external drive using the 3-2-1 rule.

- Share drone maps with your agronomist before field visits to transform the conversation from discovery to targeted diagnosis. Track recommendations against outcomes using drone and yield data.
- Shapefiles are the universal file format for prescription transfer. GeoTIFF is the standard for map imagery. Both work across John Deere, Case IH, AGCO, and aftermarket systems.
- Test the data pipeline in the off-season: create a dummy prescription, load it onto equipment, and verify the display reads it correctly before you're under in-season time pressure.
- Common integration problems (file won't load, rates are wrong, boundaries are offset) have straightforward fixes. Most trace to file format issues, coordinate system mismatches, or attribute column naming.
- The complete flight-to-prescription-to-application workflow takes 4–5 days. Start simple and build toward the full integrated workflow over 2–3 seasons.
- Any use of drone data is better than no use. A farmer who flies and makes a better flat-rate decision is getting value. The full VRT integration is the goal you grow into, not the minimum requirement.

Chapter 18 Notes and References

1. John Deere. “Operations Center: Data Management and Prescription Maps.” https://www.deere.com/en/technology-products/. Accessed February 2026.

2. Climate FieldView. “Field Data Platform: Data Import and Visualization.” https://climate.com/. Accessed February 2026.

3. Ag Leader Technology. “SMS Advanced Software: Data Management and Prescription Creation.” https://www.agleader.com/. Accessed February 2026.

4. USDA NRCS. “Web Soil Survey.” https://websoilsurvey.nrcs.usda.gov/. Accessed February 2026.

5. QGIS Project. “QGIS: A Free and Open Source Geographic Information System.” https://qgis.org/. Accessed February 2026.

6. Raven Industries. "Viper 4+: Prescription Import Guide." https://ravenind.com/. Accessed February 2026.

7. Case IH. "AFS Connect: Data Management Platform." https://www.caseih.com/. Accessed February 2026.

8. AGCO. "Fuse Smart Farming Solutions." https://www.agcocorp.com/fuse.html. Accessed February 2026.

9. Trimble Agriculture. "Farm Works Software and Connected Farm." https://agriculture.trimble.com/. Accessed February 2026.

10. Shannon, D.K., Clay, D.E., and Kitchen, N.R. (eds.). "Precision Agriculture Basics." ASA, CSSA, SSSA, 2018.

Note: Software platforms, menu navigation, and file format support change with manufacturer updates. The integration guidance in this chapter reflects products available as of early 2026. When in doubt, consult your equipment dealer or the manufacturer's current documentation for the most up-to-date import procedures.

Chapter 19: Common Mistakes and How to Avoid Them

Every mistake in this chapter is one I've either made myself or watched someone else make. Some cost money. Some cost time. A few cost drones. All of them are avoidable if you know to look for them. Consider this chapter the shortcut that lets you skip the expensive lessons and get straight to the productive ones.

The farmers who get the most value from drones aren't the ones who never make mistakes. They're the ones who make each mistake only once. This chapter catalogs the most common errors across four categories—equipment, flying, interpretation, and business—so you can learn from other people's experience instead of your own.

Some of these mistakes are obvious in hindsight. Others are subtle and can persist for an entire season before you realize you've been collecting bad data or making decisions on flawed assumptions. Read this chapter before your first season, and revisit it mid-season to check yourself.

Equipment Mistakes

Mistake #1: Buying Too Much Too Soon

The mistake: Spending $8,000–$15,000 on a multispectral system, enterprise drone, and premium software before you've flown a single crop flight. The salesperson or YouTube channel convinced you that nothing less than a Matrice with a MicaSense sensor is worth owning.

Why it happens: Enthusiasm outpaces experience. The advanced equipment is genuinely capable, and the marketing makes it look essential. But the gap between "capable" and "necessary for your operation right now" is enormous.

The fix: Start with a $1,200–$1,500 consumer RGB drone (Chapter 2). Fly it for one full season. At the end of the season, you'll know exactly what you need and don't need—based on your fields, your management style, and your actual use cases. Upgrade from a position of knowledge, not hope.

The cost of this mistake: $3,000–$10,000 in unnecessary equipment that either sits unused or duplicates what a simpler system could do.

Mistake #2: Skipping the Practice

The mistake: Taking the drone out of the box on the first day the crop is at V4, attempting a mapping flight over a 160-acre field, and discovering mid-flight that you don't know how to handle a low-battery return, adjust the camera settings, or recover from a lost GPS signal.

Why it happens: The growing season waits for no one. The crop is at the right stage NOW, and you don't want to waste the window practicing over an empty field.

The fix: Budget 5–10 hours of practice before your first crop flight. Chapter 11's Weeks 2–4 are designed for this. Practice flights don't need to be over agricultural land – your yard, a park, or an empty lot works fine for learning the controls. The automated mapping practice should be over an actual field, but it doesn't need to have a crop in it.

Mistake #3: Ignoring Battery Health

The mistake: Flying batteries until they won't hold a charge, storing them fully charged (or fully depleted) for months, or not tracking charge cycles. Then wondering why your 35-minute battery only lasts 22 minutes mid-season.

The fix: Store batteries at 40–60% charge when not in use for more than a week. Don't leave them fully charged for extended periods—this accelerates cell degradation. Most DJI batteries will auto-discharge to storage level after a few days if you enable the setting. Track total charge cycles; plan to replace batteries every 200–300 cycles or when capacity drops below 80% of original.

Mistake #4: Flying in Bad Conditions

The mistake: Launching in 20+ mph wind because "this is the only day I have," flying in light rain because "it's barely drizzling," or flying into the golden hour because you started the field too late. The resulting data is blurry, inconsistently exposed, or stitched poorly because the drone was fighting wind gusts.

The fix: Hard limits: no flying above 15 mph sustained wind, no rain, and complete the flight before solar angles create long shadows (generally before 3:00 PM in summer). A wasted trip to the field costs you an hour. A wasted flight over the field costs you an hour plus the time to discover the data is unusable. Bad data that you don't realize is bad and act on can cost far more.

Flying Mistakes

Mistake #5: Inconsistent Flight Altitude

The mistake: Flying one field at 200 feet, the next at 300, and the third at 150 because you didn't set a standard—or because you let the mapping app default to different altitudes. The resulting maps have different resolutions and can't be directly compared.

Why it matters: Altitude determines ground resolution (Chapter 8). If you're comparing NDRE values between fields or between flights on the same field, inconsistent altitude introduces a variable that confuses interpretation. A pixel at 200 feet covers different ground area than a pixel at 300 feet, and the canopy-to-soil ratio in each pixel changes accordingly.

The fix: Standardize at 200–250 feet AGL for all agricultural monitoring flights. Save it as your default in the mapping app. Only deviate from this altitude for specific purposes (lower for stand counts, higher for very large fields).

Mistake #6: Not Enough Overlap

The mistake: Setting overlap too low (below 65% front, 55% side) to fly faster and cover more acres. The result: gaps in the orthomosaic, blurry zones where stitching failed, and areas with no usable data—often right over the area you needed to see most.

The fix: Minimum 75% front overlap, 65% side overlap for agricultural mapping. These settings produce reliable stitching for flat terrain. Don't reduce them to save time; the 5–10 extra minutes of flight time are trivial compared to discovering after processing that you have holes in your map. If processing fails or produces artifacts, increase overlap to 80/70 and refly.

Mistake #7: Flying at the Wrong Time of Day

The mistake: Flying early morning or late afternoon when shadows are long. The shadows create false dark bands across the orthomosaic and confuse vegetation index calculations. NDVI and NDRE values in shadowed areas are artificially depressed, creating phantom stress zones that don't exist.

The fix: Fly between 10:00 AM and 2:00 PM for multispectral. For RGB-only, the window extends to about 10:00 AM–4:00 PM. Overcast days (diffuse light, no shadows) actually produce excellent RGB data and are acceptable for multispectral if you're using a calibration panel. Avoid partly cloudy conditions—alternating sun and shadow across the field creates inconsistent exposure.

Mistake #8: Flying Too Often

The mistake: Flying every field every week because you can. This generates mountains of data that you don't have time to process, interpret, or act on. By mid-season, you're three flights behind on processing, and the data from the critical V10 flight is sitting unreviewed on an SD card.

The fix: Chapter 13's "few flights" philosophy: 3–4 strategic flights per field per season, timed to growth stages where decisions are imminent. Process and interpret each flight before the next one. A flight that produces a reviewed, ground-truthed, acted-upon map is worth infinitely more than five flights that pile up in a folder.

Interpretation Mistakes

Mistake #9: Acting on Bad Data Without Ground-Truthing

THE MOST EXPENSIVE MISTAKE IN THIS BOOK

Spraying a fungicide, applying extra nitrogen, or replanting based on what the map shows – without walking the field to confirm what's actually happening on the ground. A low-NDRE zone could be nitrogen deficiency, compaction, drainage, disease, hybrid boundary, or a cloud shadow during the flight. Each requires a different response, and some require no response at all.

Rule: Never spend more than $500 on a management action triggered by a drone map without first confirming the diagnosis on foot. The 30 minutes of ground-truthing can save you thousands in misdirected inputs.

Mistake #10: Ignoring Context – The Variety Boundary Trap

The mistake: Seeing a sharp line on the NDRE map and assuming it's a stress boundary, when it's actually the line between two corn hybrids or two soybean varieties that have naturally different canopy characteristics. Different hybrids have different leaf angles, chlorophyll content, and canopy architecture—all of which produce NDRE differences that have nothing to do with stress.

How to spot it: The "stress" boundary is perfectly straight and runs exactly along your planting rows. It appears at the same location on every flight throughout the season. It corresponds to where you switched seed lots or varieties in the planter. Real stress boundaries are rarely perfectly straight, and they change over time.

The fix: Keep your planting map. Overlay it with the drone map. Mark hybrid boundaries on your first map of the season and mentally adjust for them on every subsequent map. Compare NDRE values within each hybrid, not across hybrid boundaries.

Mistake #11: Expecting Too Much Precision

The mistake: Treating the NDRE map as an absolute measurement of nitrogen status, canopy health, or yield potential. Assigning dollar values to individual pixels. Drawing management zones to the nearest tenth of an acre. Arguing that a 0.02 difference in NDRE between two areas is meaningful.

The reality: Vegetation indices are relative measures influenced by dozens of variables: sun angle, atmospheric conditions, canopy moisture, soil background, sensor calibration, and processing parameters. A 0.02 NDRE difference is within the noise. A 0.10 NDRE difference across a persistent zone is meaningful. Learn to distinguish signal from noise by looking for large, persistent, spatially coherent patterns and ignoring pixel-level variation.

The fix: Use three broad management zones (Chapter 16), not twenty micro-zones. Look for patterns that persist across multiple flights, not anomalies that appear once. When in doubt, ground-truth. The map is a guide, not a GPS coordinate for where to stand.

Mistake #12: Chasing Every Anomaly

The mistake: Spending three hours walking a field to investigate every small color variation on the orthomosaic. Most of those variations are within normal range – soil texture changes, minor moisture differences, or processing artifacts. By the time you've walked every "spot," you've used a full morning and found nothing actionable.

The fix: Triage. Focus your ground-truthing on patterns, not spots. Patterns that affect 5+ acres, that are new since the last flight, that correspond to known risk zones (low drainage, high disease pressure), or that suggest a cause with an available intervention. Ignore isolated pixels, narrow artifacts along stitching boundaries, and variations smaller than a couple of acres unless they're in a critical area.

WHEN TO GROUND-TRUTH: DECISION RULES

- The pattern is NEW (not present on the previous flight)
- The pattern covers 5+ contiguous acres
- The pattern is in a management-responsive zone (not a rock outcrop or permanent waterway)
- You have a specific action you could take if you confirmed the cause (spray, side-dress, replant)
- The crop is still within the window where intervention would improve yield

  If fewer than three of these apply, the anomaly probably isn't worth investigating this week. Note it for future reference and move on.

Business Mistakes

Mistake #13: Not Calculating ROI

The mistake: Flying all season, generating maps, and feeling good about the technology—but never actually calculating whether the drone-informed decisions generated more value than they cost. After three seasons, you couldn't tell a skeptical spouse or partner whether the investment paid off because you never tracked it.

The fix: Track every management decision the drone influenced and assign a conservative value estimate. The season summary template from Chapter 11 is designed for this. Even rough estimates ("the replant decision saved about $3,000; the spot-spray saved about $1,500") provide accountability. If the math doesn't work after two seasons, you either need to change how you're using the data or acknowledge that the technology isn't the right fit for your operation.

Mistake #14: Collecting Data You Won't Use

The mistake: Flying every field, generating every index (NDVI, NDRE, GNDVI, SAVI, EVI, OSAVI, CIgreen), processing at maximum resolution, and archiving everything – without a specific management question driving the data collection. The result is terabytes of beautifully processed maps that no one looks at.

The fix: Before every flight, ask: "What question am I trying to answer?" If the answer is "I don't know, I just want to fly," postpone the flight until you have a purpose. Every flight should have a specific agronomic question (stand assessment, nitrogen evaluation, disease scouting, drainage check) that determines what you fly, what index you process, and what action you'll take based on the result.

Mistake #15: Over-Investing Before Proving Value

The mistake: Buying a multispectral sensor, subscribing to premium processing software, purchasing a dedicated field laptop, and signing up for a precision ag training course – all before your first crop flight. Total investment: $8,000–12,000 before generating a single dollar of return.

The fix: The staircase approach from Chapter 2. Start with the $1,200–$1,500 RGB drone and free or low-cost processing software. Prove value in Season 1 (most farmers find $2,000–$10,000 in first-year ROI from stand assessment, replant decisions, and scouting efficiency alone). Upgrade only when you've identified a specific need that your current equipment can't meet. Every dollar of upgrade should be justified by a specific management decision that the upgrade enables.

Red Flags Checklist: Signs Your Data May Be Bad

Before acting on any drone map, scan for these red flags. If you find one, investigate before making management decisions.

Red Flag	What It Looks Like	Likely Cause	What to Do
Streaks or bands across the map	Parallel lines of different color/brightness	Inconsistent exposure; shadows; partly cloudy during flight	Refly on a uniform-light day
Blur or smearing in portions of the map	Loss of detail in sections; "soft" areas	Wind gusts; camera vibration; shutter speed too slow	Check camera settings; fly in calmer conditions
Gaps or holes in the orthomosaic	Missing sections; black areas; distorted edges	Insufficient overlap; stitching failure; GPS dropout	Refly with higher overlap (80/70); check GPS signal
Sharp straight lines in index map	Abrupt color change along a straight boundary	Hybrid/variety boundary; flight line seam; stitching artifact	Check planting map; compare to RGB; look for processing seam
Index values at extremes (0 or 1.0+)	Pixels reading as maximum or minimum	Soil pixels; water pixels; sensor saturation; dead plants	Exclude non-crop pixels; check for standing water or bare soil
Dramatically different results from previous flight	A field that looked healthy now shows severe stress everywhere	Different time of day; different weather; sensor calibration drift	Compare flight conditions between the two dates; recalibrate
Color cast across entire map	Everything shifted blue, yellow, or another hue	White balance error; no calibration panel used	Recalibrate; use reflectance calibration panel (multispectral)
Results that contradict what you see on the ground (or vice versa)	Map says healthy, but the crop is visibly stressed	Cloud shadow during flight; processing error; wrong index	Walk the field; compare map to ground reality; reprocess

Annual Self-Assessment: Are You Getting Better?

Answer these questions at the end of each season. They're designed to track whether your drone program is improving year over year—not just in data collection, but in the management value the data produces.

Question	Year 1	Year 2	Year 3
How many management decisions did drone data directly influence?	___	___	___
Estimated total value of those decisions ($)	___	___	___
How many flights produced data I actually used?	___	___	___
How many flights were wasted (bad data, no action taken)?	___	___	___
Did I ground-truth every area before taking action?	___	___	___
Did I compare maps between flights to track change?	___	___	___
Did I overlay drone maps with yield data at year-end?	___	___	___

Question	Year 1	Year 2	Year 3
What was my biggest interpretive mistake this season?	—	—	—
Am I processing maps within 48 hours of the flight?	—	—	—
Is my data organized and backed up?	—	—	—
Am I sharing maps with my agronomist before field visits?	—	—	—
What specific upgrade or change would most improve my program next year?	—	—	—

A healthy trend looks like this: the number of influenced decisions increases while the number of wasted flights decreases. The estimated value grows as your interpretation skills sharpen and you learn which flights and which fields deliver the highest return. If Year 3 looks the same as Year 1, something in your workflow needs to change.

Chapter Summary

- Equipment mistakes: don't buy too much too soon ($1,200–$1,500 RGB first), don't skip practice (5–10 hours before crop flights), maintain battery health (40–60% storage charge), and don't fly in bad conditions (15 mph wind limit, no rain, proper time of day).
- Flying mistakes: standardize altitude (200–250 feet), maintain sufficient overlap (75/65% minimum), fly during midday hours (10

AM–2 PM for multispectral), and resist the urge to fly too often (3–4 strategic flights beat 15 random ones).

- The most expensive interpretation mistake is acting on drone data without ground-truthing. Never commit more than $500 to a management action triggered by a map without walking the field first.
- Watch for the variety-boundary trap: sharp straight lines on NDRE maps often reflect hybrid differences, not stress. Keep your planting map and overlay it with every drone map.
- Vegetation indices are relative measures, not absolute. A 0.02 NDRE difference is noise; a 0.10 difference across a persistent zone is signal. Use broad zones (3–5), not micro-zones.
- Triage your ground-truthing: focus on patterns that are new, larger than 5 acres, in management-responsive areas, and within the window for intervention. Ignore isolated pixels and one-time artifacts.
- Business mistakes: track ROI explicitly (assign conservative dollar values to each decision), fly with a specific question in mind, and upgrade equipment only when justified by a specific need the current system can't meet.
- Scan every map for red flags before acting: streaks, blur, gaps, sharp straight lines, extreme values, dramatic changes from previous flights, color casts, or contradictions with ground observations.
- Complete the annual self-assessment at season's end. A healthy drone program shows increasing management decisions influenced and decreasing wasted flights over time.

Chapter 19 Notes and References

1. Mulla, D.J. "Twenty-Five Years of Remote Sensing in Precision Agriculture: Key Advances and Remaining Knowledge Gaps." Biosystems Engineering, 114(4), 358–371, 2013.

2. Hunt, E.R. and Daughtry, C.S.T. "What Good Are Unmanned Aircraft Systems for Agricultural Remote Sensing and Precision Agriculture?" International Journal of

Remote Sensing, 39(15–16), 5345–5376, 2018.

3. DJI. "Intelligent Flight Battery: Care and Storage Guide." https://www.dji.com/. Accessed February 2026.

4. Pix4D. "Image Quality and Acquisition: Best Practices." https://support.pix4d.com/. Accessed February 2026.

5. DroneDeploy. "Common Mapping Errors and How to Fix Them." https://www.dronedeploy.com/. Accessed February 2026.

6. Maes, W.H. and Steppe, K. "Perspectives for Remote Sensing with Unmanned Aerial Vehicles in Precision Agriculture." Trends in Plant Science, 24(2), 152–164, 2019.

7. Chapters 2, 8, 9, 10, 11, 13, 14, 15, 19, and 21 of this book provide the detailed best practices referenced in this chapter's mistake-avoidance guidance.

Note: The mistakes documented in this chapter are drawn from

conversations with hundreds of agricultural drone operators, university extension specialists, and service providers across the Midwest. The cost estimates for each mistake are conservative; actual losses from misdirected management actions can be significantly higher. When in doubt, ground-truth.

Epilogue

What's Next: The Future of Farm Drones

Predictions about agricultural technology have a spotty track record. GPS guidance exceeded everyone's expectations. Precision ag profitability took decades longer than promised. The "smart farm" of 2015 was supposed to be here by now. Some of it is. Most of it isn't. This epilogue tries to sort the likely from the unlikely—knowing that some of these predictions will look foolish in five years.

Every technology chapter in a book risks becoming outdated the day it's printed. Drone technology is evolving faster than most agricultural tools, and anything specific written about upcoming products has a limited shelf life. So instead of reviewing specific products (which will change), this epilogue focuses on the underlying trends (which are more durable) and provides a framework for evaluating new claims as they appear.

I've organized the future into three tiers based on how confident I am that the technology will deliver practical value for Midwest row crop operations: coming soon (high confidence, 1–3 years), on the horizon (moderate confidence, 3–7 years), and probably hype (low confidence for row crops, at least in the foreseeable future). I've also included a section on what won't change—because in a discussion about the future, the most useful predictions may be about the constants.

Where the Technology Is Headed

Coming Soon: High Confidence (1–3 Years)

These improvements are already in development or follow predictable engineering trajectories. They'll make drone monitoring easier and more accessible, but they won't fundamentally change what you do – they'll make the current workflow faster, cheaper, and more reliable.

Better battery life. Battery energy density is improving at roughly 5–8% per year, driven by massive investment in electric vehicles. Within 1–3 years, expect consumer agricultural drones to deliver 40–50 minutes of flight time on a single charge – enough to map 200–300 acres without a battery swap. This matters because battery limitations currently force multi-battery missions on larger fields, adding landing, swapping, and relaunching time. Longer flights simplify operations and reduce the temptation to cut overlap to squeeze more acreage from a single charge.

Lower-cost multispectral options. The price gap between RGB and multispectral is narrowing. In 2020, the cheapest multispectral setup cost $8,000–$12,000. By 2026, entry-level systems are approaching $3,000–$5,000. The trend will continue as sensor manufacturing scales and competition increases. Within 1–3 years, expect capable multispectral systems under $3,000 – potentially integrated into consumer platforms rather than requiring separate enterprise drones. When multispectral reaches the $2,000–$3,000 price point, it moves from an "advanced upgrade" to a "serious starting option" for any farmer monitoring more than a few hundred acres of corn. This shifts the entire economics of Chapter 7's investment decision.

Improved AI for automated analysis. The biggest near-term change in agricultural drone technology isn't in the hardware—it's in the software. Machine learning and AI-based image analysis are rapidly improving at automating tasks that currently require human interpretation: counting plants for stand assessment, identifying weed species from RGB imagery, detecting disease symptoms in canopy photos, and classifying management zones from index data. DroneDeploy, Pix4Dfields, Sentera, and several startup companies already offer AI-assisted features. Within 1–3 years, expect these tools to reliably automate plant counting, generate preliminary weed maps from RGB, and flag anomalies for ground-truthing. The emphasis is on "preliminary" and "flag"—AI assists your interpretation, it doesn't replace your judgment.

Easier integration with farm management software. The compatibility friction described in Chapter 18—getting drone data into your tractor display, your farm management platform, and your agronomist's hands—will ease significantly. John Deere, Climate FieldView, AGCO Fuse, and others are actively building drone data import features. Within 1–3 years, expect one-click import of drone maps and prescriptions into the major farm management platforms, wireless sync between drone software and in-cab displays, and standardized data formats that eliminate most file conversion issues.

On the Horizon: Moderate Confidence (3–7 Years)

These technologies exist in research, prototype, or early commercial stages. They'll likely reach practical agricultural use within the next decade, but the timeline, cost, and specific form factor are uncertain.

Autonomous "always watching" drones. Several companies are developing drone-in-a-box systems: a drone that lives in a weatherproof charging station at the field's edge, automatically launches on a schedule or trigger, flies a pre-programmed route, collects data, returns to the box, uploads to the cloud, and recharges. No human pilot required. The technology works in demonstrations. The challenges are regulatory (current FAA rules require a remote pilot to maintain visual line of sight or obtain specific waivers), economic ($20,000–$50,000+ per system today), and practical (thunderstorms mid-flight, hawk encounters, charging station power loss). These are solvable, and the FAA is moving toward expanded autonomous operations, but widespread adoption is 3–7 years away for row crop agriculture.

Real-time disease and pest identification. Research labs have demonstrated machine learning models that can identify tar spot, gray leaf spot, SDS, and several other diseases from high-resolution RGB images taken at close range. The models recognize visual characteristics of disease lesions—color, shape, texture, pattern—in ways that approach expert human identification. The gap between lab demonstration and field-ready tool is real: models trained on controlled images struggle with variable lighting, mixed pathology, overlapping symptoms, and the diversity of what a camera sees at canopy level. But within 3–7 years, expect drone software to offer preliminary disease identification from close-range RGB—not perfect, not a substitute for a pathologist, but good enough to say "this pattern is consistent with tar spot; ground-truth to confirm" rather than just "something is wrong here."

Drone-to-robot integration. The next logical step after drone-guided prescriptions: the drone identifies a weed patch and an autonomous ground sprayer treats it within hours – no human driving required. John Deere's autonomous tractor program, Carbon Robotics' laser weeding system, and several startups are building the individual components. Integrating them into a seamless drone-to-robot pipeline is an engineering and data interoperability challenge, not science fiction. The 3–7 year timeline assumes continued progress on autonomous vehicle regulation and farm-scale robotics reliability.

Probably Hype for Row Crops

Some technologies generate headlines and trade show excitement but face fundamental challenges that make them unlikely to deliver practical value for Midwest row crop operations in the foreseeable future.

Drone-based pollination makes no agronomic sense for corn (wind-pollinated) or soybeans (self-pollinating). The concept has some relevance for specialty fruit crops where pollinator decline threatens production, but even there, the economics are challenging and the biology is complex. For row crops, this is not a technology you need to follow.

Individual plant management from drones sounds impressive but collides with physics and economics. Current drone sensors at practical flight altitudes cannot resolve individual corn plants reliably enough for per-plant management. Even if resolution improves, the application equipment to deliver variable inputs at the individual-plant level doesn't exist for field crops. Zone-based management at the 1–5 acre level captures the vast majority of manageable variability.

Full automation of scouting misunderstands what scouting is. Scouting isn't just looking at the crop from above. It's pulling back husks to check ear development. It's digging roots to assess compaction. It's splitting stems to look for disease. It's feeling the soil. It's noticing that the field smells different after a rain. None of this can be done from 200 feet in the air, regardless of sensor technology. Drones will continue to improve as scouting multipliers – showing you where to walk and what to look for. But they won't eliminate the need to walk.

Technology Maturity Assessment

Technology	Maturity	Timeline	Value for Row Crop Farmers	Action
Longer battery life (40–50 min)	Near-production	1–2 years	High	Wait for next purchase cycle; don't buy now expecting current specs
Sub-$3,000 multispectral	Early commercial	1–3 years	High	Monitor pricing; plan upgrade when price/value aligns
AI-assisted plant counting	Commercial (early)	Available now–2 years	Moderate–High	Try current offerings in DroneDeploy, Pix4Dfields
AI-assisted weed ID (RGB)	Late research / early commercial	2–4 years	Moderate	Watch; don't pay premium yet

Technology	Maturity	Timeline	Value for Row Crop Farmers	Action
One-click farm software integration	In development	1–3 years	High	Provide feedback to your platform vendor; drives development
Drone-in-a-box (autonomous)	Prototype / early commercial	3–7 years	Moderate	Monitor; economics don't work for most farms yet
AI disease identification	Late research	3–7 years	Moderate–High	Interesting but not reliable enough to act on without ground-truthing
Drone-to-robot integration	Early research / prototype	5–7+ years	Potentially high	Long-term watch; not actionable now
Swarm operations	Prototype	5–7+ years	Low (for family farms)	Relevant mainly for service providers
Individual plant management	Conceptual / early research	10+ years (if ever)	Low for row crops	Skip; zone management captures available value
Full scouting automation	Conceptual	Not foreseeable	Low	Skip; boots on the ground remain essential

What Won't Change

In a discussion about the future, the most useful predictions may be about what stays the same. These constants are worth more than any technology forecast because you can build your management philosophy around them with confidence.

Agronomic knowledge will always be required to interpret data. No AI, no software, no algorithm can replace the farmer who understands why a particular soil type on a particular field behaves the way it does. Technology produces data; people produce understanding. The farmer who combines drone data with 30 years of field-specific experience will always outperform the farmer who relies on technology alone.

Ground-truthing will always be necessary. Remote sensing – whether from drones, satellites, or future autonomous platforms – will always see the surface and the aggregate. The specific diagnosis – the root cause of what's happening below the canopy, within the plant, or in the soil – requires direct observation. The technology that tells you where to look will keep improving. The need to look will not go away.

Experience and observation are irreplaceable. The most sophisticated sensor system cannot replicate the judgment that comes from decades of watching crops grow on the same land. A drone map is a single snapshot of a dynamic, living system. The farmer's accumulated observation—spanning seasons, weather cycles, management experiments, and the stories passed down from previous generations—provides context that no data set can contain.

Boots on the ground will always matter. This book has argued throughout that drones are most valuable as complements to, not substitutes for, traditional farming skills. That argument only strengthens as technology advances. The more powerful the technology becomes, the more important the human judgment that interprets it, validates it, and decides what to do with it. The future of farm drones is not a future without farmers. It's a future where farmers see more, know more, and make better decisions—with their boots still on the ground.

How to Evaluate New Technology Claims

As drone technology continues to evolve, you'll encounter marketing claims from manufacturers, software companies, and service providers. Use this framework to separate genuine value from sales pitch.

FIVE QUESTIONS TO ASK ABOUT ANY NEW AG DRONE TECHNOLOGY

1. What specific management decision does this help me make better? If the answer is vague ("better data," "more information"), be skeptical. Good technology answers a specific question.

2. What does it cost per acre, including my time? Total cost matters, not just equipment price. Include learning time, processing time, and the opportunity cost of your hours.

3. Has it been validated on Midwest row crops by independent researchers? University trials or published results carry more weight than manufacturer demos. Ask for references.

4. Can I try it on a small scale before committing? A product that requires a $10,000 commitment before you can evaluate it on your own fields is a higher risk than one you can trial on 40 acres.

5. Does it integrate with what I already have, or does it require a new ecosystem? The best technology fits into your existing workflow. Beware products that require replacing your entire data management system.

These five questions won't make every decision easy, but they'll keep you grounded in practical value rather than technological excitement. The history of precision agriculture is littered with products that generated impressive data but didn't change management decisions – and therefore didn't generate returns. The technology that matters is the technology that helps you farm better.

Resources for Staying Current

Drone technology will continue to evolve after this book is printed. These free and low-cost resources will help you stay informed without drinking from the marketing firehose.

Resource	Type	What It Covers	Cost
Ohio State C.O.R.N. Newsletter	Weekly email (in-season)	Crop conditions, pest alerts, technology updates for Ohio	Free
Purdue Pest & Crop Newsletter	Weekly email (in-season)	Indiana-focused crop management and technology	Free
Precision Ag magazine / website	Monthly / online	Industry-wide precision ag news and product reviews	Free online
CropLife / Farm Equipment magazines	Monthly / online	Equipment, technology, and retailer perspectives	Free online
InfoAg Conference	Annual conference	Leading precision ag conference with drone sessions	Registration fee
Farm Science Review (Ohio)	Annual outdoor expo	Equipment demos including drones; OSU Extension presentations	Admission fee
DroneDeploy / Pix4D blogs	Online (ongoing)	Software updates, new features, agricultural use cases	Free

Resource	Type	What It Covers	Cost
University extension websites (OSU, Purdue, Illinois)	Online (ongoing)	Research results, recommendations, technology evaluations	Free
YouTube: agricultural drone channels	Online (ongoing)	Flight tutorials, product reviews, field demonstrations	Free (quality varies)

Final Thoughts

Nineteen chapters ago, we started with a simple premise: drones are tools that help farmers see their fields more clearly, make better decisions, and manage their acres more profitably. Nothing in the intervening pages—the science, the operations, the applications, the advanced capabilities, or the future possibilities—changes that premise.

The drone doesn't farm for you. It doesn't replace your experience, your judgment, or your daily attention to the land. What it does is extend your vision and compress your scouting time, giving you the information to act on what you already know how to do—grow crops, manage inputs, and steward the land.

Start small. Fly with purpose. Ground-truth what you see. Act on what you confirm. Review what happened. Adjust and improve.

Eyes in the sky. Boots on the ground. That's the whole program.

Notes and References

1. Tsouros, D.C., Bibi, S., and Sarigiannidis, P.G. "A Review on drone-Based Applications for Precision Agriculture." Information, 10(11), 349, 2019.

2. Rejeb, A., Abdollahi, A., Rejeb, K., and Treiblmaier, H. "Drones in Agriculture: A Review and Bibliometric Analysis." Computers and Electronics in Agriculture, 198, 107017, 2022.

3. Kim, J., Kim, S., Ju, C., and Son, H.I. "Unmanned Aerial Vehicles in Agriculture: A Review of Perspective of Platform, Control, and Applications." IEEE Access, 7, 105100–105115, 2019.

4. Federal Aviation Administration. "Advanced Aviation Advisory Committee (AAAC): Beyond Visual Line of Sight Operations." FAA Report, 2024.

5. John Deere. "Autonomous Tractor and Integrated Precision Agriculture Vision." https://www.deere.com/. Accessed February 2026.

6. DroneDeploy. "AI and Machine Learning in Agricultural Analytics: Product Roadmap." https://www.dronedeploy.com/. Accessed February 2026.

7. Ohio State University. "C.O.R.N. Newsletter." https://agcrops.osu.edu/newsletter/corn-newsletter. Accessed February 2026.

8. Purdue University. "Pest & Crop Newsletter." https://extension.entm.purdue.edu/pestcrop/. Accessed February 2026.

Note: Technology predictions in this epilogue reflect the author's assessment as of early 2026. The agricultural drone industry is evolving rapidly, and some predictions will prove optimistic, others conservative. The framework for evaluating new technology (the five questions) will remain useful regardless of which specific predictions prove accurate.

Appendixes A–I

These appendixes provide the practical tools, reference materials, and templates you need to put the book's concepts into action. Appendix A offers printable quick-reference cards for field use. Appendix B matches equipment recommendations to your budget. Appendix C walks through software tutorials step by step. Appendix D lists Ohio and Midwest resources. Appendix E provides a comprehensive 109-term glossary. Appendix F is a troubleshooting guide for common problems. Appendix G includes ROI worksheets. Appendix H covers thermal imaging for water stress detection. Appendix I provides flight log and record templates.

Appendix A: Quick Reference Cards

These reference cards are designed to be printed and laminated for field use. Each fits on a single page.

Pre-Flight Checklist

PRE-FLIGHT CHECKLIST—PRINT AND LAMINATE

BEFORE LEAVING HOME

[ ] Check weather: wind < 15 mph, no rain, no storms forecast

[ ] Charge all batteries (drone, controller, phone/tablet)

[ ] Clear/format SD card; verify storage space

[ ] Check LAANC/airspace (B4UFLY app or FAA DroneZone)

[ ] Pack: drone, controller, batteries, landing pad, phone/tablet, charger, sunshade

AT THE FIELD

[ ] Select launch site: flat, dry, away from power lines and trees

[ ] Deploy landing pad

[ ] Inspect propellers for damage, cracks, or debris

[ ] Verify battery charge level (drone and controller)

[ ] Power on controller, then drone; wait for GPS lock

[ ] Verify return-to-home altitude is set above obstacles

[ ] Check camera settings (exposure, white balance)

[ ] Set flight plan in mapping app (altitude, overlap, boundary)

[ ] Visual scan: no aircraft, no people in flight area

AFTER FLIGHT

[ ] Verify images on SD card (spot-check for sharpness)

[ ] Power off drone, then controller

[ ] Record flight in log (date, field, conditions, notes)

[ ] Transfer images to computer; begin processing same day

Vegetation Index Interpretation Guide

Index	Full Name	What It Measures	Range	When to Use	Key Thresholds
NDVI	Normalized Difference Vegetation Index	General plant vigor and green biomass	0 to 1.0	Early season; canopy closure; general health	<0.4 = sparse/stressed; 0.4–0.7 = moderate; >0.7 = dense/healthy
NDRE	Normalized Difference Red-Edge	Chlorophyll content; nitrogen status	0 to ~0.6	V8–R2 for nitrogen; mid-season health	<0.25 = low (investigate); 0.25–0.40 = moderate; >0.40 = healthy
GNDVI	Green Normalized Difference Vegetation Index	Chlorophyll concentration; similar to NDVI	0 to 1.0	Alternative to NDVI; dense canopy	Similar to NDVI; slightly more sensitive to chlorophyll
SAVI	Soil-Adjusted Vegetation Index	Vegetation vigor with soil correction	-1 to 1.0	Early season when soil visible through sparse canopy	Corrects NDVI inflation from soil background; use pre-canopy closure

Growth Stage Timing Chart

>

Growth Stage	Approx. Timing (Central OH)	Recommended Camera	Key Decisions
V2–V4	2–3 weeks after planting	RGB	Replant? Planter adjustments?
V6–V8	4–6 weeks after planting	RGB	Spot-spray weeds? Baseline established?
V8–V10 (critical)	6–8 weeks after planting	Multispectral preferred	Side-dress rate? VRT zones?
VT–R2	8–11 weeks after planting	Either	Fungicide? Scout priorities?
R4–R5	12–14 weeks after planting	RGB	Harvest timing? Stalk integrity?

Growth Stage Timing Chart—Soybeans

Growth Stage	Approx. Timing	Recommended Camera	Key Decisions
V3–V4	3–4 weeks after planting	RGB	Stand adequate? Early weeds?
R2–R3	8–10 weeks after planting	Either	Fungicide for white mold? SDS?
R5–R6	12–14 weeks after planting	RGB	Harvest sequence? Late disease?

Emergency Procedures

EMERGENCY QUICK REFERENCE

Lost connection / flyaway:

Do NOT panic. Most drones auto-return-to-home on signal loss. Wait 30 seconds. If drone does not return, note last known position and direction.

Low battery warning:

Immediately initiate return-to-home. Do not try to finish the mission. Land at the nearest safe location if RTH won't make it back.

Aircraft in the area:

Immediately descend to below 100 feet and hold position or land. Yield right-of-way to all manned aircraft. Resume only when airspace is clear.

Erratic flight behavior:

Land immediately using manual controls. Do not attempt return-to-home if the drone is not flying predictably. Power off and inspect before flying again.

Crash or hard landing:

Power off immediately. Inspect for propeller damage, arm cracks, gimbal damage, and battery integrity. Do NOT fly again until all components are verified undamaged.

Appendix B: Equipment Recommendations by Budget

Prices reflect early 2026 market. Check current availability before purchasing.

Table B.1: Equipment Recommendations by Budget Tier

Budget Tier	Drone	Key Specs	Software	Best For
$500–$1,000	DJI Mini 3 / Mini SE	12 MP RGB; 25–30 min flight; <249g	DroneDeploy (free tier)	Learning; basic scouting; limited mapping
$1,000–$2,000 (Recommended Start)	DJI Mini 4 Pro or Air 3	48 MP RGB; 34–46 min flight; obstacle avoidance	DroneDeploy or Pix4Dcapture	Stand counts, weed scouting, drainage, visual health
$2,000–$5,000	DJI Air 3 + Sentera NDVI	RGB drone + add-on single NDVI sensor	Pix4Dfields ($1,650/yr)	All RGB applications + basic vegetation index mapping
$5,000–$10,000	DJI Matrice 350 + MicaSense RedEdge-P	5-band multispectral; 42 min flight; enterprise-grade	Pix4Dfields	Full NDRE nitrogen management, VRT prescriptions
$10,000+	DJI M350 + MicaSense Altum-PT or Sentera 6X	Multispectral + thermal + high-res RGB in one sensor	Pix4Dfields + QGIS	Full precision agriculture: NDRE + thermal + VRT

Essential accessories for any tier: extra SD cards (128 GB), spare propellers, landing pad ($15–$20), tablet sunshade, hard carrying case (if not included), external hard drive for backup.

Appendix C: Software Tutorials (Step-by-Step)

DroneDeploy: Basic Mapping Flight

Step	Action	Notes
1	Open DroneDeploy app; connect to drone	Ensure drone is powered on and GPS has lock
2	Tap "New Plan"; select your field on the satellite map	Drag corners to match field boundary
3	Set altitude: 200–250 feet	Higher = faster coverage; lower = higher resolution
4	Set overlap: 75% front, 65% side	Don't reduce below these minimums
5	Verify flight path covers entire field (no gaps at edges)	Extend boundary slightly beyond field edges
6	Tap "Fly"; confirm takeoff	Drone flies automatically; monitor progress on screen
7	After landing, images auto-upload to DroneDeploy cloud	Wi-Fi required; processing takes 2–24 hours depending on field size

QGIS: Viewing and Analyzing Drone Maps (Free)

Step	Action	Notes
1	Download QGIS from qgis.org (free); install	Works on Windows, Mac, Linux

Step	Action	Notes
2	Open QGIS; drag your GeoTIFF orthomosaic into the map canvas	File automatically georeferenced
3	To add soil map: download shapefile from Web Soil Survey; drag into QGIS	Both layers align by GPS coordinates
4	To add yield map: import shapefile from combine/farm software; drag into QGIS	Adjust transparency to overlay layers
5	Use "Identify" tool (i icon) to click any point and read values	Shows NDRE value, soil type, yield—whatever layer is selected
6	To create zones: Raster > Raster Calculator; reclassify NDRE into three classes	Use thresholds from Chapter 16: <0.35, 0.35–0.45, >0.45
7	Convert zones to shapefile: Raster > Conversion > Polygonize	Creates vector zones you can edit and export
8	Export prescription shapefile: right-click layer > Export > Save Features As	Save as ESRI Shapefile (.shp); this is the format most displays accept
9	Transfer via USB drive to display terminal or upload to operations center platform	Verify column name matches what your display expects (usually "Rate" or "VR_RATE")

Creating Prescription Files for Equipment

Brand	Preferred Format	Import Method	Key Notes
John Deere (Gen 4)	Shapefile (.shp)	Operations Center web > Import > assign to field > sync to display	Rate column must be named correctly; WGS84 projection
Case IH (AFS)	Shapefile (.shp)	USB to AFS Pro display > Import Prescription	Same file as JD; verify column naming
AGCO (Fuse)	Shapefile or ISO-XML	Fuse platform web > Import > sync to display	ISO-XML for newer ISOBUS systems
Raven Viper 4	Shapefile (.shp)	USB > Import Prescription > straightforward import	Widely compatible; match rate column
Ag Leader InCommand	Shapefile (.shp)	USB > Field > Import > select prescription file	SMS Advanced software can transfer wirelessly
Trimble FMX/TMX	Shapefile (.shp)	USB > File Manager > Import Prescription	Farm Works software recommended for prescription creation

Appendix D: Ohio and Midwest Resources

University Extension Programs

Institution	Program	Contact / URL
Ohio State University Extension	Precision Agriculture Program; C.O.R.N. Newsletter	https://agcrops.osu.edu 614-292-6891
Purdue University Extension	Digital Agriculture; Pest & Crop Newsletter	https://ag.purdue.edu 765-494-8491
University of Illinois Extension	Precision Conservation Management; farmdoc	https://extension.illinois.ed u 217-333-2660
University of Wisconsin Extension	Tarspotter; Sporecaster; UAS Research	https://ipcm.wisc.edu
University of Kentucky Extension	Precision Agriculture and Drone Research	https://www.uky.edu/Ag

Conferences and Events

Event	When	Where	What to Expect
Farm Science Review	September (annual)	London, Ohio	Equipment demos, Extension presentations, drone exhibits
InfoAg Conference	July (annual)	Varies (Midwest)	Leading precision ag conference; drone sessions and workshops

Event	When	Where	What to Expect
Ohio Agri-Business Association	January–February	Columbus, Ohio	Industry conference; precision ag technology track
Indiana CCA Conference	December (annual)	Indianapolis, IN	Continuing ed for CCAs; precision ag and drone content
Commodity Classic	February–March (annual)	Varies (National)	National farm show; drone exhibits and precision ag

Online Communities and Forums

Community	Platform	Focus
DroneDeploy Community Forum	Web (community.dronedeploy.com)	Software support, agricultural use cases, user tips
r/drones and r/farming	Reddit	General drone discussion and farming technology
Agricultural Drones Facebook groups	Facebook	User-to-user advice, equipment reviews, field examples
Ag Data Transparent	Web (agdatatransparent.com)	Data ownership and privacy in precision agriculture
Precision Ag Alliance	Various	Industry organization connecting farmers with precision ag providers

Appendix E: Glossary of Terms

109 Terms – Expanded Edition

This expanded glossary includes all technical terms, acronyms, and concepts used throughout the book. Terms are defined in the context of agricultural drone operations for Midwest row crop farming. Cross-references to relevant chapters are included where applicable.

A

Aerial Application. The spraying or distribution of agricultural chemicals (pesticides, herbicides, fertilizers) from aircraft, including drones. Requires Part 137 certification.

AGL (Above Ground Level). Altitude measurement from the ground surface directly below the drone, as opposed to above sea level (ASL). Most agricultural flights operate between 200-400 feet AGL.

Autonomous Flight. Drone operation following a pre-programmed flight path without continuous manual control input, typically managed through flight planning software.

B

B4UFLY. FAA mobile application that provides real-time airspace information, helping drone pilots determine whether it is safe and legal to fly in a specific location.

Band (Spectral Band). A specific range of wavelengths in the electromagnetic spectrum that a camera sensor captures. Common agricultural bands include red, green, blue, red edge, and near-infrared.

Biomass. The total quantity of plant material in a given area, often estimated through vegetation indices derived from drone imagery.

C

Calibration Panel. A reference target with known reflectance values placed on the ground before flights to calibrate multispectral camera readings for consistent, comparable data.

Canopy. The uppermost layer of crop foliage in a field. Canopy temperature, color, and density are key indicators captured by drone sensors.

Canopy Temperature. The temperature of the crop canopy surface measured by thermal cameras. Elevated canopy temperature indicates water stress as transpiration decreases.

CEC (Cation Exchange Capacity). A soil property that affects nutrient availability and can contribute to spatial variability visible in drone imagery.

Change Detection. The practice of comparing drone maps from different dates to identify temporal changes in crop health, growth, or field conditions.

Chlorophyll. The green pigment in plants responsible for photosynthesis. Chlorophyll content strongly influences how plants reflect light, particularly in red and red-edge wavelengths.

Cloud Processing. Using remote servers (cloud computing) to process drone imagery into maps, as opposed to processing on a local computer. Services include DroneDeploy and Pix4D Cloud.

Controller (Drone). The handheld device used to manually fly the drone and monitor its status, camera feed, and telemetry data during flight.

Controller (Equipment). The onboard computer on agricultural equipment (sprayers, planners, fertilizer applicators) that reads prescription maps and varies application rates accordingly.

Cross-Reference. A citation or mention within one chapter directing the reader to related content in another chapter of the book.

CWSI (Crop Water Stress Index). A thermal-derived index that quantifies the degree of water stress a crop is experiencing by comparing actual canopy temperature to reference temperatures.

D

DEM (Digital Elevation Model). A 3D representation of terrain elevation generated from drone imagery, useful for identifying drainage patterns and topographic features.

Digital Surface Model (DSM). A 3D model that includes the height of all surface features including vegetation and structures, as opposed to a DEM which represents bare ground.

DJI. Da-Jiang Innovations, the world's largest consumer and commercial drone manufacturer. Products discussed include the Mini series, Mavic 3 Multispectral, and Agras spray drones.

DroneDeploy. A cloud-based drone mapping and analytics platform commonly used in agriculture for processing orthomosaics and vegetation index maps.

DroneZone (FAA). The FAA's online portal where drone pilots register their aircraft and apply for airspace authorizations.

DSM (Digital Surface Model). See Digital Surface Model.

E

EAHV (Excess Apparent Heat Variation). A thermal analysis metric that identifies areas with unusually high canopy temperatures, potentially indicating water stress or disease.

Electromagnetic Spectrum. The full range of electromagnetic radiation, from radio waves to gamma rays. Agricultural remote sensing primarily uses visible light (400-700 nm) and near-infrared (700-1300 nm).

Emergence. The stage when crop seedlings first break through the soil surface. Early post-emergence flights are valuable for stand count assessment.

EVI (Enhanced Vegetation Index). A vegetation index that improves on NDVI by correcting for atmospheric conditions and soil background effects. Particularly useful in high-biomass conditions.

ExG (Excess Green). An RGB-based vegetation index that highlights green vegetation against soil and residue backgrounds. Useful with standard cameras for early-season plant counting.

F

FAA (Federal Aviation Administration). The United States government agency responsible for regulating civil aviation, including all drone operations under 14 CFR Part 107.

Field of View (FOV). The angular extent of the observable area that the drone camera can capture in a single image. Wider FOV covers more ground per image but with less resolution.

Flight Log. A record of each drone flight documenting date, time, location, weather conditions, altitude, battery usage, and observations. Recommended for legal protection and operational improvement.

Flight Plan. A pre-programmed route the drone follows during autonomous operation, specifying waypoints, altitude, speed, and camera settings for consistent data collection.

G

GCI (Green Chlorophyll Index). A vegetation index sensitive to chlorophyll content, calculated using green and near-infrared reflectance bands.

Georeferenced. Data that has been tagged with geographic coordinates (latitude and longitude), allowing it to be accurately placed on a map and overlaid with other spatial data.

GeoTIFF. A standard image file format that includes embedded geographic coordinate information, commonly used for drone-derived orthomosaics and index maps.

GPS (Global Positioning System). Satellite-based navigation system used by drones for positioning, navigation, and geotagging images. RTK

GPS provides centimeter-level accuracy.

Ground Control Points (GCPs). Physical markers placed in the field with precisely surveyed GPS coordinates, used to improve the geometric accuracy of drone-generated maps.

Ground Resolution. See Ground Sampling Distance (GSD).

Ground Sampling Distance (GSD). The distance between the centers of two consecutive pixels measured on the ground. Lower GSD means higher spatial resolution. At 250 feet AGL, typical GSD is 0.5-1.5 inches per pixel.

Ground-Truthing. The practice of physically visiting locations identified as anomalous on drone maps to verify what the aerial imagery shows and confirm the cause of observed patterns.

H

Heading. The compass direction the drone's nose points during flight. Consistent heading during mapping ensures uniform sun angle effects across images.

Heatmap. A color-coded visualization where different colors represent different values (e.g., NDVI scores, temperature readings), making spatial patterns immediately visible.

Hybrid Seed. Seed produced by cross-pollinating two different inbred parent lines. Hybrid vigor and maturity dates are important factors in replant decisions.

I

Image Overlap. The percentage of shared area between adjacent drone photographs. Standard agricultural mapping requires 75-80% frontal overlap and 65-70% side overlap for reliable stitching.

Index Map. A map generated by applying a vegetation index formula (such as NDVI) to multispectral drone imagery, displaying crop health or other parameters as color-coded values.

ISOXML. An ISO standard data format used by precision agriculture equipment for exchanging prescription maps and task data between software and equipment controllers.

K

L

LAANC (Low Altitude Authorization and Notification Capability). An FAA system that provides automated airspace authorization for drone operations in controlled airspace near airports, typically through apps like AirMap or Aloft.

LiDAR (Light Detection and Ranging). A remote sensing technology that uses laser pulses to measure distances and create 3D models. Increasingly available on agricultural drones for terrain and canopy mapping.

M

Management Zone. A subdivision of a field that receives uniform treatment based on shared characteristics identified through drone imagery, yield data, soil sampling, or other data sources.

Mavic 3 Multispectral (DJI). A DJI drone platform that combines an RGB camera with a four-band multispectral camera (green, red, red edge, NIR) designed specifically for agricultural monitoring.

MicaSense. A manufacturer of multispectral camera systems commonly used in agricultural drone applications, including the RedEdge and Altum series.

Mini Series (DJI). DJI's sub-249-gram consumer drone line (Mini 2, Mini 3 Pro, Mini 4 Pro) popular as entry-level agricultural scouting platforms due to their portability and low cost.

Mosaic. See Orthomosaic.

Multispectral Camera. A camera that captures images in multiple specific wavelength bands simultaneously, including bands beyond human vision such as near-infrared and red edge.

N

NDRE (Normalized Difference Red Edge). A vegetation index calculated from red edge and near-infrared bands. More sensitive than NDVI to nitrogen content and chlorophyll levels in mid-to-late season crops.

NDVI (Normalized Difference Vegetation Index). The most widely used vegetation index, calculated as (NIR - Red) / (NIR + Red). Values range from -1 to 1, with healthy vegetation typically scoring 0.6-0.9.

Near-Infrared (NIR). Electromagnetic radiation with wavelengths between approximately 700-1300 nm, just beyond visible red light. Healthy plants strongly reflect NIR, making it crucial for vegetation health assessment.

Nitrogen (N). An essential plant macronutrient that is typically the largest variable input cost in corn production. Nitrogen deficiency causes yellowing (chlorosis) and reduced yields.

O

Orthomosaic. A geometrically corrected, georeferenced image created by stitching together hundreds of overlapping aerial photographs into a single seamless map. Overlap. See Image Overlap.

P

Part 107. The section of Federal Aviation Regulations (14 CFR Part 107) governing commercial small unmanned aircraft systems (sUAS) operations, including all agricultural drone use for business purposes.

Part 137. The section of Federal Aviation Regulations governing agricultural aircraft operations, including aerial application of pesticides and fertilizers by drone.

Part 44809. The section of U.S. Code establishing rules for recreational drone operations, also known as the Exception for Limited Recreational Operations.

Payload. The weight a drone carries beyond its own structure, including cameras, sensors, batteries, and (for spray drones) liquid tanks.

Photogrammetry. The science of making measurements from photographs. In drone agriculture, photogrammetry software stitches overlapping images into orthomosaics and 3D models.

Photosynthesis. The process by which plants convert light energy, water, and carbon dioxide into glucose and oxygen. Photosynthetic activity directly affects how plants reflect light.

Pix4D. A photogrammetry software company offering Pix4Dfields (agricultural mapping), Pix4Dmapper (general mapping), and cloud processing services.

Pixel. The smallest unit of a digital image. In drone imagery, each pixel corresponds to a specific ground area (see Ground Sampling Distance).

Plant Count. See Stand Count.

Precision Agriculture. A farming management approach that uses technology to observe, measure, and respond to spatial and temporal variability in crops and fields.

Prescription Map. A digital file containing spatially variable application rates for inputs like nitrogen, seed, or pesticides, generated from drone data and loaded onto equipment controllers.

Q

QGIS. A free, open-source geographic information system (GIS) software used for viewing, editing, and analyzing geospatial data including drone-generated maps.

R

Radiometric Calibration. The process of converting raw sensor values to actual reflectance values using calibration panels, enabling consistent comparisons across flights and dates.

Red Edge. The spectral region between approximately 680-730 nm where plant reflectance transitions sharply from low (red absorption) to high (NIR reflectance). Changes in the red edge indicate chlorophyll and nitrogen status.

Reflectance. The proportion of incoming light that is reflected by a surface. Different plant conditions produce different reflectance patterns across the electromagnetic spectrum.

Remote ID. An FAA requirement (effective 2024) that drones broadcast identification and location information during flight, similar to a digital license plate.

Remote Pilot Certificate. The FAA certification required under Part 107 for commercial drone operations, obtained by passing the Unmanned Aircraft General knowledge exam.

Remote Sensing. The acquisition of information about an object or area from a distance, typically using sensors on satellites, aircraft, or drones.

Replant Decision. The determination of whether to replant a field or area based on stand count data, date, hybrid maturity, and economic analysis. Drone imagery aids this critical early-season decision.

RGB (Red, Green, Blue). The three primary color channels captured by standard cameras. RGB imagery is useful for visual scouting, stand counts, and some vegetation analyses.

ROI (Return on Investment). A measure of the profitability of an investment. In agricultural drone context, ROI compares the cost of drone technology against savings in inputs, time, or increased yields. RTK (Real-Time Kinematic). A GPS correction technique that provides centimeter-level positioning accuracy. RTK-equipped drones produce more geometrically accurate maps.

S

SAVI (Soil-Adjusted Vegetation Index). A vegetation index that minimizes the influence of soil brightness on the vegetation signal, particularly useful for early-season mapping when canopy cover is sparse.

Scouting. The practice of physically inspecting crop fields to assess plant health, pest presence, weed pressure, and other conditions. Drone imagery enhances scouting efficiency by directing where to look.

Sentera. A manufacturer of agricultural drone sensors and analytics platforms, offering both standalone sensors and integrated drone systems.

Service Provider. A professional drone operator hired to conduct flights, process data, and deliver maps and analytics to farmers who prefer not to operate their own equipment.

Shapefile. A common geospatial data format used for sharing prescription maps, management zones, and field boundaries between GIS and farm management software.

Sidedress. A method of applying nitrogen fertilizer alongside growing crop rows during the growing season, typically at V6-V8 growth stage in corn. Drone data helps optimize sidedress rates.

Spectral Signature. The unique pattern of light reflectance across different wavelengths produced by a specific material or condition. Healthy and stressed plants have distinct spectral signatures.

Stand Count. The number of plants per unit area in a field, assessed to evaluate planting success and inform replant decisions. Drone imagery enables rapid, accurate stand counts over large areas. Stitching. The process of combining multiple overlapping photographs into a single seamless image (orthomosaic) using photogrammetry software.

Swarm Technology. The coordinated operation of multiple drones simultaneously, potentially allowing faster field coverage. Currently in development for agricultural applications.

T

Thermal Camera. A sensor that captures infrared radiation to measure surface temperatures. In agriculture, thermal cameras detect canopy temperature variations indicating water stress.

Tile Drainage. Underground perforated pipes installed to remove excess water from agricultural fields. Thermal and RGB drone imagery can help identify drainage problems and map tile line locations.

Transpiration. The process by which plants release water vapor through leaf pores (stomata). Reduced transpiration raises canopy temperature, detectable by thermal imaging.

TSAVI (Transformed Soil-Adjusted Vegetation Index). An advanced version of SAVI that further minimizes soil background effects, useful in fields with variable soil exposure.

U

UAS (Unmanned Aircraft System). The complete system including the drone aircraft, controller, communication links, and any associated support equipment. The FAA's formal term for drone systems.

V

Variable Rate Application (VRA). The practice of applying different rates of inputs (fertilizer, seed, chemicals) to different areas of a field based on prescription maps derived from drone or other data.

Variable Rate Technology (VRT). Equipment and software systems that enable variable rate application, including GPS-guided controllers, flow meters, and prescription map software.

Vegetation Index. A mathematical formula applied to spectral reflectance data that highlights vegetation properties such as health, density, or chlorophyll content. Common indices include NDVI, NDRE, and SAVI.

VLOS (Visual Line of Sight). The FAA requirement that drone pilots maintain the ability to see their aircraft with unaided vision during the entire flight. This limits practical flight distance.

VRA. See Variable Rate Application.

VRT. See Variable Rate Technology.

W

Water Stress. A condition where crop water demand exceeds supply, causing reduced transpiration, elevated canopy temperature, and eventually visible wilting. Detectable by thermal drone imaging.

Waypoint. A predetermined GPS coordinate that defines a point along the drone's flight path during autonomous operation.

WebODM. A free, open-source photogrammetry platform for processing drone imagery into orthomosaics, digital elevation models, and point clouds.

Y

Yield Map. A georeferenced map showing crop yield variation across a field, typically generated from combine harvester data. Yield maps are valuable for validating drone-based management decisions.

Z

Zone Map. See Management Zone.

Note: Terms specific to multispectral imaging, variable-rate technology, and FAA regulations are defined in the context most relevant to Midwest row crop agriculture. For terms with broader industry usage, the definition provided here reflects the meaning used in this book.

Appendix F: Troubleshooting Guide

Problem	Possible Cause / Solution
Drone won't connect to controller	Update firmware; try different USB cable; check cable for damage; move away from power lines and towers
Drone won't acquire GPS lock	Move to open area; remove from metal surfaces (magnetic interference); wait 2–3 minutes
Images are blurry	Fly in calmer winds (under 15 mph); set shutter to 1/1000s if able; clean lens before flight
NDVI/NDRE values look wrong	Use calibration panel before each flight; verify correct band assignments in software

Appendix G: ROI Worksheets and Calculators

Equipment Cost Recovery Calculator

Item	Your Cost
Drone purchase price	$_______
Extra batteries	$_______
Accessories (SD cards, case, landing pad)	$_______
Software subscription (annual)	$_______
Total Year 1 investment (A)	$_______
Annual maintenance / battery replacement (B)	$_______
Annual time investment: _____ hours x $_____ /hr value (C)	$_______
Total annual cost (Year 1: A+B+C; Year 2+: B+C)	$_______

Input Savings Estimator

Savings Category	Calculation	Your Estimate
Nitrogen savings (VRT vs. flat rate)	Acres saved x lbs N saved x $/lb N	$_______
Replant decision (avoided unnecessary replant)	Acres NOT replanted x seed + labor cost/acre	$_______
Spot spray vs. whole-field treatment	(Whole-field cost) – (spot-spray cost)	$_______

Savings Category	Calculation	Your Estimate
Fungicide targeting (treated only needed areas)	Acres NOT treated x fungicide + application $/acre	$________
Seed savings (VR population)	Acres reduced x seeds/acre reduced x $/seed	$________
Drainage repair prioritization	Value of correctly targeted repair vs. guesswork	$________
Scouting time saved	Hours saved x $/hour value	$________
Total annual savings (D)	$________	

Net ROI

Annual ROI Calculation

Net annual benefit = Total savings (D) – Total annual cost (A+B+C Year 1; B+C in Year 2+)

Payback period = Total Year 1 investment (A) ÷ Annual net benefit

Your net annual benefit: $________ Your payback period: ________ seasons

Appendix H: Thermal Imaging for Water Stress Detection

Thermal Imaging—Water Stress and Beyond

A healthy plant is a cool plant. When roots can't supply enough water, the stomata close, transpiration stops, and the leaf temperature rises. A thermal camera turns that invisible temperature difference into a map – showing you water stress before the crop shows any visible sign of wilting.

RGB cameras see light. Multispectral cameras see chlorophyll. Thermal cameras see heat. Each sensor type adds a layer of information that the others can't provide, and thermal imaging fills a specific gap: it measures the crop's water status directly, through the physical relationship between plant water use and leaf temperature.

For most dryland Midwest operations, thermal imaging is an interesting but nonessential addition. The management decisions it drives – primarily irrigation scheduling – don't apply to rainfed fields. But for irrigated operations, for fields with significant drainage variability, and as a diagnostic tool that complements multispectral data, thermal imaging offers genuine, measurable value.

This chapter explains the science behind thermal crop sensing, walks through the primary agricultural applications, provides honest guidance on when the investment makes sense, and gives you a practical framework for interpreting thermal maps.

How Thermal Imaging Works

The Plant-Temperature Principle

Every object emits thermal radiation – infrared energy proportional to its temperature. A thermal camera detects this radiation and converts it into a temperature map. In agriculture, the object of interest is the crop canopy, and the temperature it measures reflects the plant's physiological status.

The mechanism is elegantly simple. A healthy, well-watered plant keeps its stomata open, allowing transpiration—the evaporation of water from leaf surfaces. Transpiration cools the leaf, just as sweating cools your skin. A well-watered cornfield at midday in July might have a canopy temperature of 80–85°F even when the air temperature is 90–95°F. The crop is actively cooling itself through water use.

When the plant runs low on water—whether from drought, root restriction, or insufficient irrigation—it closes its stomata to conserve moisture. Transpiration slows or stops. Without the cooling effect of evaporation, the leaf temperature rises toward ambient air temperature and sometimes above it. A water-stressed corn canopy at the same time and place might register 92–98°F.

That 7–15°F difference between well-watered and stressed canopy is what thermal cameras detect. From 250 feet up, this temperature difference is mapped across the entire field, showing you exactly which zones are under water stress and which are adequately supplied.

THE THERMAL LOGIC

- *Well-watered plant: Stomata open -> Transpiration active -> Leaf cooled by evaporation -> Canopy temperature BELOW air temperature*
- *Water-stressed plant: Stomata closing -> Transpiration reduced -> Less evaporative cooling -> Canopy temperature NEAR air temperature*
- *Severely stressed plant: Stomata closed -> No transpiration -> No evaporative cooling -> Canopy temperature ABOVE air temperature*

What the Camera Sees

A thermal camera produces a grayscale or false-color image where each pixel represents a temperature value. Cool areas appear as one color (typically blue or dark) and warm areas as another (typically red or bright). The resulting map looks similar to an NDVI map in structure – a color-coded field image showing spatial variability – but the data represents temperature rather than vegetation index values.

Most agricultural thermal cameras have a resolution of 160x120 to 640x512 pixels and measure temperatures in the range of -4°F to 1,112°F with sensitivity (Noise Equivalent Temperature Difference, or NETD) of 0.05–0.1°F. This sensitivity means the camera can detect temperature differences of a fraction of a degree – far more precise than what you'd need for agricultural applications where the stress-versus-healthy difference is typically 5–15°F.

The Crop Water Stress Index (CWSI)

Raw canopy temperature alone is hard to interpret because it's influenced by air temperature, humidity, wind speed, and solar radiation—all of which change throughout the day and between days. The Crop Water Stress Index (CWSI) normalizes the thermal data by comparing the observed canopy temperature to the theoretical temperature of a fully-watered canopy (lower baseline) and a completely non-transpiring canopy (upper baseline) under the same environmental conditions.

CWSI ranges from 0 (no stress—canopy fully cooled by transpiration) to one (maximum stress—no transpiration). A CWSI above 0.3–0.4 typically indicates the crop is experiencing meaningful water deficit. Some thermal processing software calculates CWSI automatically; otherwise, you'll work with relative temperature differences between areas of the field.

Primary Applications of Thermal Imaging in Agriculture

Irrigation Scheduling and Management

This is the flagship application and the one where thermal imaging delivers its clearest ROI. For irrigated operations – center pivots, linear systems, or drip irrigation – thermal maps answer two critical questions: when does the crop need water, and where does it need it most?

Traditional irrigation scheduling relies on soil moisture sensors (point measurements that may not represent the whole field), weather-based evapotranspiration models (estimates based on weather data, not actual crop condition), or visual observation (by the time you see wilting, yield loss has already begun). Thermal imaging measures the crop's actual water status across the entire field simultaneously.

The decision rule is straightforward: when the thermal map shows the canopy temperature rising above the "well-watered" baseline—or CWSI exceeding 0.3—it's time to irrigate. Catching the stress signal 2–3 days before visible wilting gives you a window to schedule irrigation before yield-limiting stress sets in.

Not all areas of an irrigated field receive equal water. Center pivots apply more water at the outer towers (traveling faster, greater arc distance) and may miss corners. Uneven soil types absorb and drain water at different rates. Equipment malfunctions – a clogged nozzle, a worn gasket, an improperly aimed sprinkler head – create dry spots.

Thermal maps reveal all of these issues as patterns of elevated canopy temperature. A dry spot from a clogged nozzle appears as a warm arc corresponding to the nozzle's position on the pivot arm. Uneven soil drainage shows as warm (dry) sandy ridges and cool (wet) clay bottoms. Corner areas outside the pivot's reach are consistently warmer than the irrigated interior.

Thermal Pattern	Probable Cause	Action
Warm arc or band at consistent radius from pivot center	Clogged, damaged, or misaligned sprinkler nozzle	Inspect and repair the specific nozzle/span
Warm wedge or sector within the pivot circle	Pivot speed inconsistency or end-gun timing issue	Check pivot panel timer settings; verify end-gun timing
Warm patches on sandy soil ridges	Faster drainage on coarse-textured soil	Consider VRI (variable-rate irrigation) or soil amendment

Thermal Pattern	Probable Cause	Action
Uniformly elevated temperature across entire circle	Insufficient application rate or scheduling	Increase run time or frequency; check well capacity
Warm corners outside the pivot radius	Normal – not reached by pivot	Install corner system or accept dryland yield in corners
Cool (wet) patches in low areas	Over-irrigation or poor drainage in low spots	Reduce rate in low zones; consider surface drainage

Drainage Problem Identification

In Chapter 13, we discussed using post-rain RGB flights to identify drainage patterns. Thermal imaging adds another dimension: saturated soil and standing water are cooler than dry soil. On a warm day after a rain event, a thermal map can reveal drainage patterns even after the visible water has disappeared from the surface.

Tile lines show up in thermal imagery as subtle cool lines (the tile is draining subsurface water, keeping the soil along the tile slightly wetter and cooler than the surrounding soil). Broken tiles may show as a transition from cool (functioning upstream) to warm (failing downstream where the soil drains poorly). Surface drainage paths appear as cool channels winding through warmer upland areas.

For dryland Midwest operations, this drainage application may be the strongest reason to consider thermal imaging. The ability to map tile lines, identify breaks, and visualize subsurface drainage patterns on a warm day – without needing rain or visible surface water – adds a powerful diagnostic tool to the drainage assessment toolkit described in Chapter 13.

Crop Stress Confirmation and Diagnosis

When your NDRE map shows a low-vigor zone, thermal data can help narrow the diagnosis. Chapter 12's five-step diagnostic process asks "Is the nitrogen available but the roots can't reach it?" Thermal imagery helps answer this by indicating whether the stress is water-related.

Low NDRE + warm canopy: Likely water stress or root restriction (compaction, restricted drainage). The plant can't access enough water, so it closes stomata, reducing both transpiration (warm canopy) and photosynthesis (low chlorophyll / low NDRE).

Low NDRE + cool canopy: Likely nutrient deficiency or disease – the plant has adequate water (transpiring normally, canopy cool) but is struggling for another reason. This distinction directs your ground-truthing toward nutrient or pathogen issues rather than water or root problems.

Normal NDRE + warm canopy: Early water stress that hasn't yet affected chlorophyll. The plant is starting to close stomata but hasn't exhausted its nitrogen reserves. This is the early-warning scenario where thermal catches stress before NDRE shows it.

THE DIAGNOSTIC POWER OF COMBINING THERMAL AND MULTISPECTRAL

Used alone, NDRE tells you the crop is stressed but not why and Thermal tells you the canopy is warm but not what's causing it. Combined, they create a two-dimensional diagnostic: the NDRE axis shows chlorophyll status, the thermal axis shows water status. Together they narrow the diagnosis more efficiently than either alone.

This is an advanced technique that requires both sensor types and the ability to overlay maps. It's most valuable on fields with multiple interacting stressors – drought plus nitrogen deficiency, or compaction plus disease – where single-sensor diagnosis is ambiguous.

Scenario: Thermal Value by Operation Type

Scenario	Thermal Value	Recommendation
Irrigated row crops (pivot, linear)	High	Strong candidate – irrigation scheduling and system diagnostics provide clear ROI
Irrigated specialty crops (vegetables, orchards)	Very high	Excellent fit – high crop value justifies sensor cost; water management is critical
Dryland crops with drainage variability	Moderate	Useful for tile mapping and drainage diagnostics; seasonal rather than routine use
Dryland crops, well-drained, uniform fields	Low	Limited value – RGB and multispectral cover most management decisions
Livestock operations (not covered in this book)	Variable	Thermal has applications in animal health; outside scope of this chapter

The decision matrix is driven by one question: does water management account for a significant portion of your management decisions and costs? If yes, thermal imaging provides data that other sensors cannot. If your operation is dryland with no irrigation and relatively uniform drainage, the information thermal provides is interesting but rarely changes management decisions beyond what NDRE already tells you.

Equipment Options and Cost

Thermal Cameras for Agricultural Drones

Camera / System	Resolution	Compatibility	Approx. Cost (2026)	Notes
DJI Mavic 3T (Thermal)	640x512	DJI Enterprise platform	$4,500–$5,200	Integrated thermal + RGB; enterprise-grade; complete drone
DJI Zenmuse H20T	640x512	DJI Matrice 300/350 (payload only)	$6,500–$8,000	Multi-sensor pod; zoom + wide + thermal
FLIR Vue Pro R	336x256 or 640x512	Multiple platforms via mount	$2,500–$5,000	Radiometric; standalone thermal sensor
DJI Zenmuse XT2	640x512	DJI Matrice 200/300 (payload only)	$5,000–$7,000	FLIR sensor integrated with RGB
Sentera 6X Thermal	640x512	DJI Matrice / Sentera	$6,000–$8,000	6-band multispectral + thermal in one sensor
Budget thermal (FLIR Lepton-based)	160x120	Various via adapter	$500–$1,500	Low resolution; adequate for large-scale irrigation patterns

The most practical entry point for thermal imaging in 2026 is the DJI Mavic 3 Thermal—a complete drone system with integrated thermal and RGB cameras for approximately $4,500–$5,200. It's flight-ready out of the box and provides 640x512 thermal resolution that's more than adequate for agricultural applications. If you already own a DJI Matrice enterprise drone, adding a thermal payload is a logical upgrade.

Budget options using FLIR Lepton-based modules ($500–$1,500) provide lower resolution (160x120) that's sufficient for identifying large-scale irrigation patterns but too coarse for the detailed drainage analysis or stress diagnosis work described earlier. For most farmers, the resolution difference between budget and mid-range thermal justifies the additional investment.

Cost-Benefit Analysis

THERMAL IMAGING ROI: IRRIGATED CORN EXAMPLE

Operation: 400 irrigated acres under 3 center pivots
Thermal system cost: $4,800 (Mavic 3 Thermal)

Annual Benefits:

Water savings from precision scheduling: 1.5 inches/season x 400
Acres = 16.3 million gallons saved x $0.004/gal pumping cost = $650/year

Yield recovery from catching stress 2–3 days earlier: 3 bu/acre on 25% of acres = 300 bu x $4.50 = $1,350/year

For dryland operations using thermal primarily for drainage diagnostics, the ROI calculation is different. You're not capturing annual water savings—you're making better one-time investment decisions about tile drainage. If thermal data prevents even one misdirected tile repair ($1,500–$3,000 in avoided cost) or identifies a broken tile that's been costing you $500/year in yield, the sensor pays for itself. But the payback is lumpy and less predictable than the irrigated scenario.

Flying for Thermal Data: Timing and Conditions

Thermal data is far more sensitive to flight conditions than RGB or multispectral. The same field flown at 8:00 AM and 1:00 PM will produce dramatically different thermal maps—not because the crop changed, but because the sun angle, air temperature, and humidity changed. Getting useful, interpretable thermal data requires disciplined flight timing.

Optimal Conditions

Condition	Ideal	Acceptable	Avoid
Time of day	Solar noon ± 1 hour	10 AM–3 PM	Early morning or late afternoon
Sky	Clear, full sun (>80%)	Mostly clear	Overcast or partly cloudy
Wind	Calm (<5 mph)	Light (5–10 mph)	Moderate to strong (>10 mph)
Recent rain	None in past 24+ hours	Light rain >12 hours prior	Rain within 6 hours (wet canopy invalidates data)
Air temperature	Above 75°F	Above 65°F	Below 60°F (temperature differences compress)

The rationale for these conditions: you want maximum contrast between well-watered and stressed canopy. That contrast is greatest when solar radiation is strong (clear sky, midday), the air is warm (driving transpiration demand), and the wind is calm (wind-induced convective cooling masks the transpiration signal). A warm, clear, calm day around solar noon is your target. Overcast, cool, or windy conditions compress the temperature differences and make the data harder to interpret.

THE MIDDAY RULE

For RGB and multispectral, we recommended overcast conditions or solar noon ±2 hours.

For thermal, the recommendation is the opposite:

1. *Clear sky and solar noon ±1 hour. This means you may need to fly thermal and multispectral at different times – or accept that a*

2. *Single midday flight on a clear day works adequately for both, even though it's not ideal for multispectral radiometric precision.*

Interpreting Thermal Maps

When you review a thermal map, you're looking for relative temperature differences within the field, not absolute temperatures. The absolute canopy temperature depends on ambient conditions, but the difference between a well-watered zone and a stressed zone is diagnostically meaningful regardless of the baseline.

Temperature range of 2–4°F within the field: Normal variability. Minor differences from soil type, microclimate, or slight moisture variation. No action needed.

Temperature range of 5–8°F: Meaningful variability. The warmer zones are experiencing measurable water stress. If irrigated, prioritize these zones for the next irrigation pass. If dryland, the stress may be driven by soil or root factors worth investigating.

Temperature range of 10°F or more: Severe stress or a non-crop feature. Warm spots exceeding 10°F above the field average likely represent bare soil (exposed by stand gaps or lodging), equipment paths, or areas where the canopy has died. Ground-truth to confirm.

Unusually cool areas: Saturated soil, standing water, or areas receiving excess irrigation. These may indicate drainage problems, over-irrigation, or low spots that accumulate water.

Limitations and Honest Assessment

Thermal imaging is a genuinely useful agricultural tool, but it has real limitations that you should understand before investing.

Narrow optimal flight window: The requirement for clear sky, midday, calm conditions limits how many days per season you can collect quality thermal data. In a humid Midwest summer with frequent afternoon clouds, you may have 8–12 good thermal windows in a typical growing season. Contrast this with RGB, which works on almost any non-rainy day.

Not a standalone diagnosis: Thermal maps show temperature patterns, not causes. A warm zone could be water stress, compaction, root disease, or bare soil. Like NDRE, thermal data requires ground-truthing and works best in combination with other sensor data.

Soil interference in sparse canopy: Early in the season, before canopy closure, the thermal camera sees a mix of plant and soil temperatures. Bare soil can be dramatically warmer than the canopy on a sunny day, contaminating the plant temperature signal. Thermal imaging is most reliable after canopy closure (V8+ in corn, R1+ in soybeans).

Resolution trade-off: Thermal sensors have lower pixel resolution than RGB or multispectral cameras. At 250 feet with a 640x512 sensor, thermal pixel size is approximately 6–8 inches – adequate for irrigation management but too coarse for plant-level analysis. This is a sensor physics limitation, not a technology shortcoming.

Learning curve: Thermal data interpretation requires understanding the relationship between canopy temperature, ambient conditions, and crop physiology. It's not as intuitive as RGB ("I can see the crop") or NDRE ("green is good, red is bad"). Budget time to learn interpretation alongside the equipment investment.

Appendix H Summary

- Thermal cameras detect crop water stress by measuring canopy temperature. Well-watered plants transpire and stay cool; water-stressed plants close stomata and warm up. The difference is typically 5–15°F and detectable before visible wilting.
- Irrigation scheduling is the primary application, answering when to irrigate (canopy warming above baseline) and where to prioritize (warmer zones first). Thermal maps also diagnose irrigation system problems—clogged nozzles, speed inconsistencies, coverage gaps.
- Drainage assessment benefits from thermal even on dryland operations. Tile lines, breaks, surface drainage paths, and ponding areas are visible as temperature patterns on warm days, even without visible surface water.

- Combining thermal and multispectral data creates a two-dimensional diagnostic: NDRE shows chlorophyll status, thermal shows water status. Together they narrow stress diagnoses more efficiently than either sensor alone.
- Thermal is most valuable for irrigated operations (clear ROI from water savings and stress prevention) and operations with significant drainage variability. For dryland fields with uniform drainage, the value is limited.
- Equipment ranges from budget thermal modules ($500–$1,500) to integrated drone systems ($4,500–$5,200). The DJI Mavic 3 Thermal is the most practical entry point for most agricultural users in 2026.
- Flight conditions matter more for thermal than any other sensor type: clear sky, solar noon ±1 hour, calm wind, no recent rain, warm air temperature. Plan thermal flights around these windows.
- Limitations include a narrow optimal flight window, soil interference in sparse canopy, lower resolution than RGB/multispectral, and a steeper interpretation learning curve. Thermal supplements but does not replace other sensor types.

Appendix H Notes and References

1. Jackson, R.D., Idso, S.B., Reginato, R.J., and Pinter, P.J. "Canopy Temperature as a Crop Water Stress Indicator." Water Resources Research, 17(4), 1133–1138, 1981.

2. Idso, S.B., Jackson, R.D., Pinter, P.J., Reginato, R.J., and Hatfield, J.L. "Normalizing the Stress-Degree-Day Parameter for Environmental Variability." Agricultural Meteorology, 24, 45–55, 1981.

3. Jones, H.G. "Use of Infrared Thermometry for Estimation of Stomatal Conductance as a Possible Aid to Irrigation Scheduling." Agricultural and Forest Meteorology, 95(3), 139–149, 1999.

4. Gonzalez-Dugo, V., Zarco-Tejada, P.J., and Fereres, E. "Applying Thermal Radiometry for Irrigation Scheduling of Citrus Orchards." Agricultural Water Management, 141, 107–116, 2014.

5. DJI. "Mavic 3 Thermal (M3T) Specifications." https://enterprise.dji.com/mavic-3-enterprise. Accessed February 2026.

6. FLIR Systems. "Vue Pro R: Radiometric Thermal Camera for Drones." https://www.flir.com/products/vue-pro-r/. Accessed February 2026.

7. Sentera. "6X Thermal: Combined Multispectral and Thermal Sensor." https://sentera.com/. Accessed February 2026.

8. Colaizzi, P.D., Barnes, E.M., Clarke, T.R., et al. "Water Stress Detection under High Frequency Sprinkler Irrigation with Water Deficit Index." Journal of Irrigation and Drainage Engineering, 129(1), 36–43, 2003.

9. Ohio State University Extension. "Irrigation Scheduling for Ohio." https://extension.osu.edu/. Accessed February 2026.

10. Hatfield, J.L. "Thermal Remote Sensing of Crop Water Stress." Advances in Agronomy, Volume 172, Chapter 5, 2022.

Note: Thermal sensor specifications and prices referenced in this chapter are based on products available as of early 2026. The thermal imaging market for agricultural drones is evolving rapidly, with sensor resolution increasing and prices decreasing. Equipment costs for irrigated operations should be weighed against your specific water costs, crop values, and irrigation system characteristics.

Appendix I: Flight Log and Record Templates

Flight Log

Copy this template for each flight. Keep in your drone data folder alongside images and maps.

Field Your Entry

Date:

Field name:

Crop / Growth stage:

Drone / Camera used:

Flight altitude (AGL):

Overlap setting (front / side):

Weather (temp, wind, sky):

Time of day (start–end):

Number of batteries used:

Number of images captured:

Purpose of flight:

Observations / notes:

Processing software used:

Outputs generated:

Management action taken (if any):

Estimated value of action ($):

Annual Equipment Maintenance Log

Date	Item	Action Taken	Cost	Next Service
________	Propellers	________	$________	________ ________ Battery 1 (cycle count: ___) ________ $________ ________ ________ Battery 2 (cycle count: ___) ________ $________ ________ ________ Battery 3 (cycle count: ___) ________ $________ ________
________	Gimbal / Camera	________	$________	________
________	Firmware update	________	N/A	________
________	Calibration panel	________	$________	________
________	SD cards / storage	________	$________	________
________	Controller	________	$________	________

Observation and Action Tracker

Use this template to connect what you see to what you do—and track whether it worked.

Date	Field	Observati on (from map)	Ground-Tr uth Finding	Action Taken	Result / Outcome
______	______	______	______	______	______
______	______	______	______	______	______
______	______	______	______	______	______
______	______	______	______	______	______
______	______	______	______	______	______
______	______	______	______	______	______
______	______	______	______	______	______
______	______	______	______	______	______

End of Reference Materials

Drones Over the Back Forty

A Midwest Farmer's Guide to Seeing What Your Eyes Simply Can't

Consolidated Bibliography

Drones Over the Back Forty

188 Sources—Chicago Manual of Style, 17th Edition

This bibliography consolidates all sources cited across 19 chapters and 9 appendixes. Citations follow the Chicago Manual of Style, 17th Edition (Notes and Bibliography system). Sources include USDA publications, university extension resources, peer-reviewed research, FAA regulations, and industry reports.

A

Abendroth, L.J., Elmore, R.W., Boyer, M.J., and Marlay, S.K. "Corn Growth and Development." Iowa State University Extension, PMR 1009, 2011.

Advexure. "DJI Mavic 3 Multispectral: Top Precision Agriculture Drone in 2025?" February 2025.

Ag Leader Technology. "SMS Advanced Software: Data Management and Prescription Creation.". https://www.agleader.com/. Accessed February 2026.

Ag Leader Technology. "SMS Advanced: Prescription Creation.". https://www.agleader.com/. Accessed February 2026.

AGCO. "Fuse Smart Farming Solutions.". https://www.agcocorp.com/fuse.html. Accessed February 2026.

American Farm Bureau Federation. "Farm Succession and Transition Planning.". https://www.fb.org/. Accessed February 2026.

American Society of Agronomy. "Certified Crop Adviser Program.". https://www.certifiedcropadviser.org/. Accessed February 2026.

Association for Unmanned Vehicle Systems International (AUVSI). "State UAS Legislation: Agricultural Applications." Updated 2025. https://www.auvsi.org/. Accessed February 7, 2026.

Association of Equipment Manufacturers. "Precision Agriculture: Dealer and Service Provider Landscape." AEM Industry Report, 2024.

B

Barnes, E.M., Clarke, T.R., Richards, S.E., et al. "Coincident Detection of Crop Water Stress, Nitrogen Status and Canopy Density Using Ground-Based Multispectral Data." Proceedings of the Fifth International Conference on Precision Agriculture, 2000.

Blackmer, T.M., Schepers, J.S., Varvel, G.E., and Walter-Shea, E.A. "Nitrogen Deficiency Detection Using Reflected Shortwave Radiation from Irrigated Corn Canopies." Agronomy Journal, 88(1), 1–5, 1996.

Bullock, D.S., Boerngen, M., Tao, H., et al. "The Data-Intensive Farm Management Project: Changing Agronomic Research Through On-Farm Experimentation." Agronomy Journal, 111(6), 2681–2698, 2019.

C

Case IH. "AFS Connect: Data Management Platform.". https://www.caseih.com/. Accessed February 2026.

Chapters 2, 8, 9, 10, 11, 13, 14, 15, 19, and 21 of this book provide the detailed best practices referenced in this chapter's mistake-avoidance guidance.

Chapters 2, 8, 9, 11, 12, 13, 14, 15, and 19 of this book provide the detailed guidance referenced in this action plan.

Climate FieldView. "Field Data Platform: Data Import and Visualization.". https://climate.com/. Accessed February 2026.

Colaizzi, P.D., Barnes, E.M., Clarke, T.R., et al. "Water Stress Detection under High Frequency Sprinkler Irrigation with Water Deficit Index." Journal of Irrigation and Drainage Engineering, 129(1), 36–43, 2003.

Crop Protection Network. "Disease Loss Estimates and Forecasting Models.". https://cropprotectionnetwork.org/. Accessed February 2026.

D

Dalla Lana, F., Plewa, D.E., Phillippi, E.S., et al. "Tar Spot of Corn: A Review of Biology, Epidemiology, and Management." Plant Disease, 105(8), 2141–2154, 2021.

Datt, B. "A New Reflectance Index for Remote Sensing of Chlorophyll Content in Higher Plants." Journal of Plant Physiology, 154(1), 30–36, 1999.

Delgado, J.A. and Bausch, W.C. "Potential use of precision conservation techniques to reduce nitrate leaching in irrigated crops." Journal of Soil and Water Conservation, 60(6), 379–387. Study conducted in northeastern Colorado.

DJI Store. "DJI Mini 4 Pro Specifications." Accessed February.

DJI. "Agras T25: Agricultural Drone Specifications.". https://ag.dji.com/t25. Accessed February 2026.

DJI. "Agras T50: Agricultural Drone Specifications.". https://ag.dji.com/t50. Accessed February 2026.

DJI. "DJI Mavic 3 Multispectral.". https://enterprise.dji.com/mavic-3-m. Accessed February 2026.

DJI. "DJI Mini 4 Pro Specifications.". https://store.dji.com/product/dji-mini-4-pro. Accessed February 2026.

DJI. "DJI Mini 4 Pro User Manual." Flight performance and battery specifications. https://store.dji.com/product/dji-mini-4-pro. Accessed February 2026.

DJI. "DJI Mini 4 Pro: Setup and Flight Guide.". https://www.dji.com/mini-4-pro. Accessed February 2026.

DJI. "DJI Mini 5 Pro Specifications.". https://www.dji.com/mini-5-pro. Accessed February 2026.

DJI. "DJI Mini 5 Pro." Product launch September 2025. https://www.dji.com/mini-5-pro. Accessed February 7, 2026.

DJI. "Enterprise Partners and Service Providers.". https://enterprise.dji.com/. Accessed February 2026.

DJI. "Intelligent Flight Battery: Care and Storage Guide.". https://www.dji.com/. Accessed February 2026.

DJI. "Mavic 3 Thermal (M3T) Specifications.". https://enterprise.dji.com/mavic-3-enterprise. Accessed February 2026.

Dorrance, A.E. "Soybean Disease Management in Ohio." Ohio State University Extension. https://extension.osu.edu/. Accessed February 2026.

Dorrance, A.E., Cruz, C., Mills, D., et al. "Management of SDS of Soybean." Ohio State University Extension, 2023.

DroneDeploy. "AI and Machine Learning in Agricultural Analytics: Product Roadmap.". https://www.dronedeploy.com/. Accessed February 2026.

DroneDeploy. "Common Mapping Errors and How to Fix Them.". https://www.dronedeploy.com/. Accessed February 2026.

DroneDeploy. "Drone Mapping Software for Agriculture.". https://www.dronedeploy.com/solutions/agriculture/. Accessed February

2026.

DroneDeploy. "Getting Started: Your First Mapping Flight.". https://www.dronedeploy.com/. Accessed February 2026.

DroneDeploy. "Service Provider Network.". https://www.dronedeploy.com/. Accessed February 2026.

Dronelink. "Terrain Following for Agricultural Mapping.". https://www.dronelink.com/. Accessed February 2026.

E

EOS Data Analytics. "Crop Monitoring.". https://eos.com/products/crop-monitoring. Accessed February 2026.

European Space Agency. "Sentinel-2 Mission.". https://sentinel.esa.int/web/sentinel/missions/sentinel-2. Accessed February 2026.

F

Federal Aviation Administration. "14 CFR Part 107 – Small Unmanned Aircraft Systems.". https://www.ecfr.gov/current/title-14/chapter-I/subchapter-F/part-107. Accessed February 7, 2026.

Federal Aviation Administration. "14 CFR Part 137: Agricultural Aircraft Operations.". https://www.ecfr.gov/. Accessed February 2026.

Federal Aviation Administration. "Advanced Aviation Advisory Committee (AAAC): Beyond Visual Line of Sight Operations." FAA Report, 2024.

Federal Aviation Administration. "Advisory Circular 137-1B: Agricultural Aircraft Operations." 2024.

Federal Aviation Administration. "Airspace 101 – Rules of the Sky.". https://www.faa.gov/uas/getting_started/airspace. Accessed February 7, 2026.

Federal Aviation Administration. "Become a Drone Pilot: Remote Pilot Certificate.".

https://www.faa.gov/uas/commercial_operators/become_a_drone_pilot. Accessed February 7, 2026.

Federal Aviation Administration. "DroneZone: UAS Registration.". https://faadronezone.faa.gov/. Accessed February 2026.

Federal Aviation Administration. "FAA DroneZone.". https://faadronezone.faa.gov. Accessed February 7, 2026.

Federal Aviation Administration. "How to Register Your Drone.". https://www.faa.gov/uas/getting_started/register_drone. Accessed February 7, 2026.

Federal Aviation Administration. "Recreational Flyers & Modeler Community-Based Organizations." Updated 2024. https://www.faa.gov/uas/recreational_flyers. Accessed February 7, 2026.

Federal Aviation Administration. "Remote Identification of Unmanned Aircraft." Final Rule, effective March 16, 2024. https://www.faa.gov/uas/getting_started/remote_id. Accessed February 7, 2026.

Federal Aviation Administration. "Remote Pilot – Small Unmanned Aircraft Systems Study Guide." FAA-G-8082-22. https://www.faa.gov/regulations_policies/handbooks_manuals/aviation/remote_pilot_study_guide. Accessed February 7, 2026.

Federal Aviation Administration. "The Recreational UAS Safety Test (TRUST).". https://www.faa.gov/uas/recreational_flyers/knowledge_test_updates. Accessed February 2026.

Federal Aviation Administration. "UAS Data Exchange (LAANC).". https://www.faa.gov/uas/programs_partnerships/data_exchange. Accessed February 7, 2026.

FLIR Systems. "Vue Pro R: Radiometric Thermal Camera for Drones.". https://www.flir.com/products/vue-pro-r/. Accessed February 2026.

G

Gates, D.M., Keegan, H.J., Schleter, J.C., and Weidner, V.R. "Spectral Properties of Plants." Applied Optics, 4(1), 11–20, 1965.

Gitelson, A.A. and Merzlyak, M.N. "Remote Estimation of Chlorophyll Content in Higher Plant Leaves." International Journal of Remote Sensing, 18(12), 2691–2697, 1997. Foundational work on red-edge indices.

Gitelson, A.A., Kaufman, Y.J., Stark, R., and Rundquist, D. "Novel Algorithms for Remote Estimation of Vegetation Fraction." Remote Sensing of Environment, 80(1), 76–87, 2002. Foundational work on visible-band vegetation indices.

Gitelson, A.A., Viña, A., Arkebauer, T.J., Rundquist, D.C., Keydan, G., and Leavitt, B. "Remote Estimation of Leaf Area Index and Green Leaf Biomass in Maize Canopies." Geophysical Research Letters, 30(5), 2003.

Gitelson, A.A., Viña, A., Ciganda, V., Rundquist, D.C., and Arkebauer, T.J. "Remote Estimation of Canopy Chlorophyll Content in Crops." Geophysical Research Letters, 32(8), 2005.

Gonzalez-Dugo, V., Zarco-Tejada, P.J., and Fereres, E. "Applying Thermal Radiometry for Irrigation Scheduling of Citrus Orchards." Agricultural Water Management, 141, 107–116, 2014.

Google. "Google Earth Pro: Historical Imagery.". https://earth.google.com/. Accessed February 2026.

Google. "Google Earth Pro.". https://earth.google.com/. Accessed February 2026.

Griffin, T.W., Shockley, J.M., and Mark, T.B. "Economics of Precision Farming." In: Shannon, D.K., Clay, D.E., and Kitchen, N.R. (eds.), Precision Agriculture Basics, ASA, CSSA, SSSA, 2018.

H

Hamza, M.A. and Anderson, W.K. "Soil Compaction in Cropping Systems: A Review of the Nature, Causes, and Possible Solutions." Soil and Tillage Research, 82(2), 121–145, 2005.

Hatfield, J.L. "Thermal Remote Sensing of Crop Water Stress." Advances in Agronomy, Volume 172, Chapter 5, 2022.

Hatfield, J.L. and Prueger, J.H. "Value of Using Different Vegetative Indices to Quantify Agricultural Crop Characteristics at Different Growth Stages Under Varying Management Practices." Remote Sensing, 2(2), 562–578, 2010.

Hatfield, J.L., Gitelson, A.A., Schepers, J.S., and Walthall, C.L. "Application of Spectral Remote Sensing for Agronomic Decisions." Agronomy Journal, 100(S3), 117–131, 2008.

Heap, I. "The International Herbicide-Resistant Weed Database." www.weedscience.org.

https://store.dji.com/product/dji-mini-4-pro. Accessed February 7, 2026.

Huete, A.R. "A Soil-Adjusted Vegetation Index (SAVI)." Remote Sensing of Environment, 25(3), 295–309, 1988. The original SAVI paper.

Hunt, E.R. and Daughtry, C.S.T. "What Good Are Unmanned Aircraft Systems for Agricultural Remote Sensing and Precision Agriculture?" International Journal of Remote Sensing, 39(15–16), 5345–5376, 2018.

Hunt, E.R., Doraiswamy, P.C., McMurtrey, J.E., Daughtry, C.S.T., Perry, E.M., and Akhmedov, B. "A Visible Band Index for Remote Sensing Leaf Chlorophyll Content at the Canopy Scale." International Journal of Applied Earth Observation and Geoinformation, 21, 103–112, 2013.

I

Idso, S.B., Jackson, R.D., Pinter, P.J., Reginato, R.J., and Hatfield, J.L. "Normalizing the Stress-Degree-Day Parameter for Environmental Variability." Agricultural Meteorology, 24, 45–55, 1981.

Input costs and equipment pricing fluctuate; verify current numbers when making your own calculations. This chapter provides a framework for analysis, not a guarantee of specific financial returns..

J

Jackson, R.D., Idso, S.B., Reginato, R.J., and Pinter, P.J. "Canopy Temperature as a Crop Water Stress Indicator." Water Resources Research, 17(4), 1133–1138, 1981.

John Deere. "Autonomous Tractor and Integrated Precision Agriculture Vision.". https://www.deere.com/. Accessed February 2026.

John Deere. "Operations Center: Data Management and Prescription Maps.". https://www.deere.com/en/technology-products/. Accessed February 2026.

John Deere. "Operations Center: Prescription Maps.". https://www.deere.com/en/technology-products/precision-ag-technology/. Accessed February 2026.

Jones, H.G. "Use of Infrared Thermometry for Estimation of Stomatal Conductance as a Possible Aid to Irrigation Scheduling." Agricultural and Forest Meteorology, 95(3), 139–149, 1999.

Jones, H.G. and Vaughan, R.A. Remote Sensing of Vegetation: Principles, Techniques, and Applications. Oxford University Press, 2010.

K

Kim, J., Kim, S., Ju, C., and Son, H.I. "Unmanned Aerial Vehicles in Agriculture: A Review of Perspective of Platform, Control, and Applications." IEEE Access, 7, 105100–105115, 2019.

Knipling, E.B. "Physical and Physiological Basis for the Reflectance of Visible and Near-Infrared Radiation from Vegetation." Remote Sensing of Environment, 1(3), 155–159, 1970. The foundational paper explaining why plants reflect NIR.

Kyveryga, P.M., Blackmer, T.M., and Mueller, D.S. "Interpreting Aerial Photographs for Crop Scouting." Iowa State University Extension, PM 1871, 2009.

L

Lan, Y. and Chen, S. "Current Status and Future Trends of Precision Agricultural Aviation Technologies." International Journal of Agricultural and Biological Engineering, 11(5), 1–17, 2018.

Licht, M. and Archontoulis, S.V. "Influence of Corn Seeding Rate, Spacing, and Emergence on Grain Yield." Iowa State University Extension, 2021.

Lindsey, A.J. and Thomison, P.R. "Corn Replant Considerations." Ohio State University Extension, C.O.R.N. Newsletter, 2023.

Lindsey, L.E. and Thomison, P.R. "Ohio Agronomy Guide." Ohio State University Extension, Bulletin 472, revised 2023.

Litchi. "Mission Planning for DJI Drones.". https://flylitchi.com/. Accessed February 2026.

Lowenberg-DeBoer, J. and Erickson, B. "Setting the Record Straight on Precision Agriculture Adoption." Agronomy Journal, 111(4), 1552–1569, 2019.

M

Maes, W.H. and Steppe, K. "Perspectives for Remote Sensing with Unmanned Aerial Vehicles in Precision Agriculture." Trends in Plant Science, 24(2), 152–164, 2019.

Mahlein, A.K. "Plant Disease Detection by Imaging Sensors – Parallels and Specific Demands for Precision Agriculture and Plant Phenotyping." Plant Disease, 100(2), 241–251, 2016.

Martin, D., Woldt, W., and Laird, D. "Use of Unmanned Aerial Systems (UAS) for Agricultural Spraying in the United States." University of Nebraska–Lincoln Extension, 2023.

Mesas-Carrascosa, F.J., Torres-Sánchez, J., Clavero-Rumbao, I., et al. "Assessing Optimal Flight Parameters for Generating Accurate Multispectral Orthomosaics by UAV." Remote Sensing, 7(10), 12793–12814, 2015.

MicaSense. "RedEdge-P Multispectral Sensor.". https://micasense.com/rededge-p. Accessed February 2026.

Mueller, D.S., Wise, K.A., Dufault, N.S., et al. "Corn Yield Loss Estimates Due to Diseases in the United States and Ontario, Canada, from 2016 to 2019." Plant Health Progress, 21(4), 238–247, 2020.

Mueller, D.S., Wise, K.A., et al. "Corn Yield Loss Estimates Due to Diseases in the United States and Ontario, Canada." Plant Health Progress, 17(3), 211–222, 2016.

Mulla, D.J. "Twenty-Five Years of Remote Sensing in Precision Agriculture." Journal of Environmental Quality, 42(2), 358–377, 2013.

N

Nafziger, E.D. "Corn Replant Decisions." University of Illinois, Department of Crop Sciences, 2022.

National 4-H Council. "4-H STEM Programs: Drone Discovery.". https://4-h.org/. Accessed February 2026.

National FFA Organization. "Precision Agriculture Career Development Event.". https://www.ffa.org/. Accessed February 2026.

Nex, F. and Remondino, F. "UAV for 3D Mapping Applications: A Review." Applied Geomatics, 6(1), 1–15, 2014.

Nielsen, R.L. "Effect of Plant Spacing Variability on Corn Grain Yield." Purdue University, Agronomy Department. AGRY-91-01, revised 2001.

Nielsen, R.L. "Estimating Yield and Dollar Returns from Corn Replanting." Purdue University, Corny News Network, May 2023. https://www.agry.purdue.edu/ext/corn/news/. Accessed February 7, 2026.

McFadden, J., Njuki, E., and Griffin, T. "Precision Agriculture in the Digital Era: Recent Adoption on U.S. Farms." USDA Economic Research Service, Economic Information Bulletin No. 248. Washington, DC: USDA ERS, 2022. https://www.ers.usda.gov/publications/pub-details/?pubid=105893.

O

Ohio Department of Agriculture. "Pesticide Regulation: Unmanned Aircraft Application.". https://agri.ohio.gov/. Accessed February 2026.

Ohio Revised Code, Chapter 4561: Aeronautics. Provisions related to unmanned aircraft systems.

Ohio State University Extension. "4-H Youth Development: STEM and Agriculture Programs.". https://extension.osu.edu/. Accessed February 2026.

Ohio State University Extension. "C.O.R.N. Newsletter: Pest and Disease Updates.". https://agcrops.osu.edu/newsletter/corn-newsletter. Accessed February 2026.

Ohio State University Extension. "Corn Growth Stages and Development." Agronomy Guide. https://extension.osu.edu/. Accessed February 2026.

Ohio State University Extension. "Irrigation Scheduling for Ohio.". https://extension.osu.edu/. Accessed February 2026.

Ohio State University Extension. "Precision Agriculture and Drone Technology for Ohio Farmers.". https://extension.osu.edu/. Accessed February 2026.

Ohio State University Extension. "Precision Agriculture for Ohio Farmers.". https://extension.osu.edu/. Accessed February 2026.

Ohio State University Extension. "Precision Agriculture Resources for Ohio Farmers.". https://extension.osu.edu/. Accessed February 2026.

Ohio State University Extension. "Replant Decisions for Corn.". https://extension.osu.edu/. Accessed February 2026.

Ohio State University Extension. "Soil Compaction: Causes, Effects, and Management.". https://extension.osu.edu/. Accessed February 2026.

Ohio State University Extension. "Tri-State Fertilizer Recommendations for Corn, Soybeans, Wheat, and Alfalfa." Bulletin 974, revised 2020.

Ohio State University Extension. "Wheat Disease Management.". https://extension.osu.edu/. Accessed February 2026.

Ohio State University. "C.O.R.N. Newsletter.". https://agcrops.osu.edu/newsletter/corn-newsletter. Accessed February 2026.

OpenDroneMap. "WebODM: Drone Mapping Made Easy.". https://www.opendronemap.org/webodm/. Accessed February 2026.

OpenDroneMap. "WebODM: Open-Source Drone Mapping.". https://www.opendronemap.org/webodm. Accessed February 2026.

P

Pedersen, P. "Soybean Growth and Development." Iowa State University Extension, PM 1945, 2009.

Pix4D. "How to Create Variable Rate Application Maps." Pix4Dfields Knowledge Base. https://support.pix4d.com/. Accessed February 2026.

Pix4D. "How to Verify Image Quality for Photogrammetry." Knowledge Base. https://support.pix4d.com/. Accessed February 2026.

Pix4D. "Image Quality and Acquisition: Best Practices.". https://support.pix4d.com/. Accessed February 2026.

Pix4D. "Image Quality for Photogrammetry: Requirements and Recommendations." Knowledge Base. https://support.pix4d.com/. Accessed February 2026.

Pix4D. "Network of Professional Drone Service Providers.". https://www.pix4d.com/. Accessed February 2026.

Pix4D. "Pix4Dcapture: Flight Planning for Mapping.". https://www.pix4d.com/product/pix4dcapture. Accessed February 2026.

Pix4D. "Pix4Dfields: Agriculture Drone Mapping Software.". https://www.pix4d.com/product/pix4dfields. Accessed February 2026.

Purdue University Extension. "Ag Innovation and Youth Outreach.". https://ag.purdue.edu/. Accessed February 2026.

Purdue University Extension. "Crop Chat and Pest & Crop Newsletter.". https://extension.entm.purdue.edu/pestcrop/. Accessed February 2026.

Purdue University Extension. "Precision Agriculture Services and Providers.". https://ag.purdue.edu/. Accessed February 2026.

Purdue University Extension. "Sidedress Nitrogen Application in Corn.". https://ag.purdue.edu/. Accessed February 2026.

Purdue University Extension. "Soybean Growth and Development.". https://ag.purdue.edu/. Accessed February 2026.

Purdue University Extension. "Soybean Replant Decisions.". https://ag.purdue.edu/. Accessed February 2026.

Purdue University. "Pest & Crop Newsletter.". https://extension.entm.purdue.edu/pestcrop/. Accessed February 2026.

Q

QGIS Development Team. "QGIS Geographic Information System." Open Source Geospatial Foundation. https://qgis.org/. Accessed February 2026.

QGIS Project. "QGIS: A Free and Open Source Geographic Information System.". https://qgis.org/. Accessed February 2026.

R

Raven Industries. "Viper 4+: Prescription Import Guide.". https://ravenind.com/. Accessed February 2026.

Raven Industries. "Viper 4+: Variable Rate Application.". https://ravenind.com/. Accessed February 2026.

Rejeb, A., Abdollahi, A., Rejeb, K., and Treiblmaier, H. "Drones in Agriculture: A Review and Bibliometric Analysis." Computers and Electronics in Agriculture, 198, 107017, 2022.

Rissing, A. "Testing the Endurance of Family Farming: A Case Study of Farm Families in the Midwestern United States." Agriculture and Human Values, 36(3), 415–429, 2019.

Roth, G.W. and Yocum, J. "Use of Remote Sensing for Monitoring Crop Condition." Pennsylvania State University Extension, 2020.

S

Sawyer, J.E. and Nafziger, E.D. "Regional Approach to Making Nitrogen Fertilizer Rate Decisions for Corn." Proceedings of the North

Central Extension-Industry Soil Fertility Conference, 2005.

Scharf, P.C., Kitchen, N.R., Sudduth, K.A., et al. "Field-Scale Variability in Optimal Nitrogen Fertilizer Rate for Corn." Agronomy Journal, 97(2), 452–461, 2005.

Sentera. "6X Multispectral Sensor.". https://sentera.com/6x. Accessed February 2026.

Sentera. "6X Thermal: Combined Multispectral and Thermal Sensor.". https://sentera.com/. Accessed February 2026.

Sentera. "FieldAgent Platform.". https://sentera.com/fieldagent/. Accessed February 2026.

Shanahan, J.F., Schepers, J.S., Francis, D.D., et al. "Use of Remote-Sensing Imagery to Estimate Corn Grain Yield." Agronomy Journal, 93(3), 583–589, 2001.

Shannon, D.K., Clay, D.E., and Kitchen, N.R. (eds.). "Precision Agriculture Basics." ASA, CSSA, SSSA, 2018.

Solari, F., Shanahan, J., Ferguson, R., Schepers, J., and Gitelson, A. "Active Sensor Reflectance Measurements of Corn Nitrogen Status and Yield Potential." Agronomy Journal, 100(3), 571–579, 2008.

Solvi. "Drone Analytics for Agriculture.". https://www.solvi.ag/. Accessed February 2026.

T

Thomason, W.E., Phillips, S.B., Davis, P.H., et al. "Early Season Nitrogen Monitoring with Active Sensors in Winter Wheat." Precision Agriculture, 12, 756–768, 2011.

Thompson, L. and Puntel, L. "Optimize your Nitrogen Investment with Precision Agriculture Technologies." University of Nebraska–Lincoln CropWatch, November 2021. https://cropwatch.unl.edu/2021/optimize-your-nitrogen-investment-precision-agriculture-technologies/. Accessed February 7, 2026.

Tom's Guide. "DJI Mini 5 Pro Review." October 2025. https://www.tomsguide.com/cameras-photography/drones/dji-mini-5-pro-review.

Accessed February 7, 2026.

Torres-Sánchez, J., López-Granados, F., and Peña, J.M. "An automatic object-based method for optimal thresholding in UAV images." Computers and Electronics in Agriculture, 114, 43–52, 2015. Research on optimal flight parameters for agricultural UAV imagery.

Trimble Agriculture. "Farm Works Software and Connected Farm.". https://agriculture.trimble.com/. Accessed February 2026.

Tsouros, D.C., Bibi, S., and Sarigiannidis, P.G. "A Review on UAV-Based Applications for Precision Agriculture." Information, 10(11), 349, 2019.

Tucker, C.J. "Red and Photographic Infrared Linear Combinations for Monitoring Vegetation." Remote Sensing of Environment, 8(2), 127–150, 1979. The paper that introduced NDVI.

Turner, D., Lucieer, A., and Watson, C. "An Automated Technique for Generating Georectified Mosaics from Ultra-High Resolution Unmanned Aerial Vehicle (UAV) Imagery." Remote Sensing, 4(5), 1392–1410, 2012.

U

U.S. Congress. "Countering CCP Drones Act." H.R. 2864, 118th Congress. https://www.congress.gov/bill/118th-congress/house-bill/2864. Accessed February 7, 2026.

U.S. Government Accountability Office. "Precision Agriculture: Benefits and Challenges for Technology Adoption and Use." GAO-24-105962, January 2024. https://www.gao.gov/products/gao-24-105962. Accessed February 7, 2026.

U.S. Government Accountability Office. GAO-24-105962, 2024.

University of Illinois Extension. "Nitrogen Management for Corn: Timing, Rate, and Source." Crop Sciences. https://extension.illinois.edu/. Accessed February 2026.

University of Wisconsin–Madison. "Sporecaster: White Mold Forecasting.". https://ipcm.wisc.edu/apps/sporecaster/. Accessed February 2026.

University of Wisconsin–Madison. "Tarspotter: Tar Spot Disease Forecasting Tool.". https://tarspotter.cals.wisc.edu/. Accessed February 2026.

USDA Economic Research Service. "Aerial imagery remains mostly grounded on U.S. farms." Charts of Note, February 2023. https://www.ers.usda.gov/data-products/charts-of-note/chart-detail?chartId=107207. Accessed February 7, 2026.

USDA Economic Research Service. "Fertilizer Use and Price.". https://www.ers.usda.gov/data-products/fertilizer-use-and-price/. Accessed February 2026.

USDA Economic Research Service. "Precision Agriculture in the Digital Era: Recent Adoption on U.S. Farms." Economic Information Bulletin No. 248, February 2023. https://www.ers.usda.gov/publications/pub-details?pubid=105893. Accessed February 7, 2026.

USDA Economic Research Service. "Precision agriculture use increases with farm size and varies widely by technology." Charts of Note, December 2023. https://www.ers.usda.gov/data-products/charts-of-note/chart-detail?chartId=110550. Accessed February 7, 2026.

USDA National Agricultural Statistics Service. "Farm Producers." 2022 Census of Agriculture Highlights, February 2024. https://www.nass.usda.gov/Publications/Highlights/2024/Census22_HL_FarmProducers_FINAL.pdf. Accessed February 7, 2026.

USDA National Agricultural Statistics Service. "USDA releases 2022 Census of Agriculture data." News Release, February 13, 2024. https://www.nass.usda.gov/Newsroom/2024/02-13-2024.php. Accessed February 7, 2026.

USDA National Agricultural Statistics Service. 2022 Census of Agriculture.

USDA Natural Resources Conservation Service. "Web Soil Survey.". https://websoilsurvey.nrcs.usda.gov/. Accessed February 2026.

USDA NRCS. "Conservation Practice Standard: Cover Crop (Code 340)." National Handbook of Conservation Practices.

USDA NRCS. "Conservation Practice Standard: Subsurface Drain (Code 606)." National Handbook of Conservation Practices.

USDA NRCS. "Web Soil Survey.". https://websoilsurvey.nrcs.usda.gov/. Accessed February 2026.

USDA Risk Management Agency. "Use of Unmanned Aerial Systems (UAS) in Crop Insurance." Policy guidance, updated 2024.

V

Verify current availability, particularly given evolving U.S. import regulations affecting DJI products..

W

Widmar, N.J., Ortega, D.L., and Thompson, N.M. "Ag Technology Adoption: Generational Perspectives." Purdue University, Center for Commercial Agriculture, 2023.

Wise, K. and Mueller, D. "Are Fungicides No Longer Just for Fungi? An Analysis of Foliar Fungicide Use in Corn." APS Features, 2011.

Woebbecke, D.M., Meyer, G.E., Von Bargen, K., and Mortensen, D.A. "Color Indices for Weed Identification Under Various Soil, Residue, and Lighting Conditions." Transactions of the ASAE, 38(1), 259–269, 1995.

X

XAG. "P100 Pro Agricultural UAS.". https://www.xa.com/en/p100. Accessed February 2026.

Z

Zhang, J., Huang, Y., Pu, R., et al. "Monitoring Plant Diseases and Pests Through Remote Sensing Technology: A Review." Computers and Electronics in Agriculture, 165, 104943, 2019.

Zucker, L.A. and Brown, L.C. "Agricultural Drainage: Water Quality Impacts and Subsurface Drainage Studies in the Midwest." Ohio State University Extension Bulletin 871, 1998.

About the Author

Adam Miller grew up in Napoleon, Ohio—a small town in the heart of northwest Ohio farm country. He didn't grow up on a farm, but he grew up in farming. In high school, he spent his summers baling hay for a local farm. It was honest, hot, physical labor—the kind that gives you a permanent respect for the people who do it year-round.

In 2006, Miller enlisted in the United States Air Force and built a career as a geospatial intelligence (GEOINT) analyst. His specialty is exactly the technology that agricultural drones are built on: multispectral sensors, thermal infrared imaging, electro-optical cameras, and the analytical methods that make raw sensor data meaningful.

His career took him from Joint Base Pearl Harbor-Hickam to Kandahar Airfield, from counter-terrorism operations in Afghanistan and Uganda to technical intelligence production at the National Air and Space Intelligence Center (NASIC) at Wright-Patterson Air Force Base in Ohio. He rose to Senior Master Sergeant and earned the Joint Service Commendation Medal, multiple Air and Space Commendation Medals, and selection as the 655th ISR Wing NCO of the Year.

When he came home to Ohio, he noticed the same remote sensing technology he'd spent his career mastering appearing over farm fields—and a gap between the people selling it and the farmers who needed to understand it. He wrote this book to bridge that gap.

Miller retired from the United States Air Force in 2026 after 20 years of service. He holds a Bachelor of Science in Security Management from Southwestern College (Kansas) and an Associate of Applied Science in Intelligence Studies from the Community College of the Air Force. He lives in Ohio with his family.

www.ingramcontent.com/pod-product-compliance
Lightning Source LLC
LaVergne TN
LVHW052024200926
848481LV00005B/1581
9798996161706